Mycorrhiza - Research, Practical Application and Future Prospects

Edited by Jair Putzke

Published in London, United Kingdom

Mycorrhiza - Research, Practical Application and Future Prospects
http://dx.doi.org/10.5772/intechopen.1009717
Edited by Jair Putzke

Contributors
Adnan Khan Niazi, Ahsan Iqbal, Alice Lemos Costa, Ana Luiza Klotz Neves, Cassiane Furlan Lopes, Faqiang Zhan, Fernando Augusto Bertazzo-Silva, Flavia Helena Aires Sousa, Ghulam Mustafa, Haihua Wang, Hongmei Yang, Huifang Bao, Irfan Ali, Jair Putzke, Jorge Renato Pinheiro Velloso, Kai Lou, Laura Amália Miranda Costa, Marcos André Pinheiro Velloso, Ming Chu, Ning Wang, Rao Sohail Ahmed Khan, Rimsha Riaz, Rong Yang, Shuang Dou, Taqwa Saeed, Tatheer Alam Naqvi, Vanielle Galiano, XinXiang Niu, Xuanqi Long, Ying Wu Shi, Yun Chen, Zhao Zhang, Zhichao Meng

First published in London, United Kingdom, 2026 by IntechOpen
IntechOpen is the global imprint of INTECHOPEN LIMITED, registered in England and Wales, registration number: 11086078, 167-169 Great Portland Street, London, W1W 5PF, United Kingdom

For EU product safety concerns: IN TECH d.o.o., Prolaz Marije Krucifikse Kozulić 3, 51000 Rijeka, Croatia, info@intechopen.com or visit our website at intechopen.com.

British Library Cataloguing-in-Publication Data
A catalogue record for this book is available from the British Library

Mycorrhiza - Research, Practical Application and Future Prospects
Edited by Jair Putzke
p. cm.
Print ISBN 978-1-83635-884-8
Online ISBN 978-1-83635-883-1
eBook (PDF) ISBN 978-1-83635-885-5

Meet the editor

Prof. Jair Putzke holds a bachelor's degree in Biological Sciences from the University of Santa Cruz do Sul (1990), a master's degree in Cryptogams from the Federal University of Pernambuco (1994), and a Ph.D. in Botany from the Federal University of Rio Grande do Sul (2003). He is currently a full professor at the Federal University of Pampa - São Gabriel campus, RS. He has experience in Botany with an emphasis on Mycology, working primarily on the following topics: fungi/bryophyte relationships, mycology, lichens, taxonomy, myxomycetes, mycorrhizae, ecology, teaching, and environmental education. He has published approximately 149 scientific articles, 28 books, and 20 book chapters. In mycology, he has mainly studied in Brazil and Antarctica, where he has worked continuously for 38 years on the relationship between fungi and bryophytes.

Contents

Preface

Mycorrhizae are a world apart: the fungi involved have microscopic vegetative bodies, generally buried in the soil, rarely emerging as macroscopic reproductive structures, and form symbiotic associations with plants. In a world where human actions govern the fate of species, they are highly threatened, especially because much of agriculture prevents them from properly associating with cultivated plants, particularly due to the widespread use of agrochemicals. Soil fertilization and the use of fungicides limit their occurrence, but they could be an important aid in improving production, especially in plants whose leaves are used commercially. Pollution in general and the use of fire, among other threats, are also affecting these species.

As if that were not enough, mycorrhizal fungi also face the environmental crisis caused by climate change. They are experts in this area, as they already enable associated plants to survive environmental and biotic stresses, potentially becoming icons in coping with these new conditions. Plants in desert and poor-soil environments rely on this alliance to survive. Many species, including Bryophytes, rely on this symbiotic relationship to survive in icy environments such as the Arctic and Antarctic.

It is still too early to predict the full effects and how fungi will react to the predicted climate changes, as a result of the meteorological system's response to the presence of excessive carbon in the atmosphere, but this is only due to a lack of studies and experts who could pay more attention to them.

This book, dedicated to gathering and updating existing knowledge and providing easy access to those interested in and beginning their scientific careers, presents reviews of various topics, allowing interested parties to quickly update their knowledge, drawing on the experiences of scientists from around the world.

In addition to the introductory chapter, four chapters with this focus have been compiled, as follows:

- The chapter "Mycorrhizal Fungi: Research Advances, Applications, and Future Perspectives" presents a review of the most comprehensive research, including molecular studies, their applications, and future perspectives, including an assessment of the impact of climate change on the group. The chapter also highlights the importance of a global database dealing with mycorrhiza and its use in agriculture.

- The chapter "Progress of Research on the Effects of Arbuscular Mycorrhizal Fungi on Plant–Soil Systems and Application of Mycorrhizal Fungi" presents a review of the progress in research and application of arbuscular mycorrhizal fungi, indicating that future research should focus on the mechanisms of plant/fungal interaction and on optimizing application technology to maximize their potential use in a sustainable agriculture and in environmental protection.

- The chapter "Advanced Bioinformatics Explores Mycological Composition and Function from Metatranscriptomic Sequencing" explores the world of meta-transcriptomic sequencing, discussing the technique and outlining its major challenges and potential advances.

- The chapter "Novel Biotechnological Methods for the Cultivation and Inoculation of Arbuscular Mycorrhizal Fungi" discusses recent advances in encapsulation and stabilization, specifically with the use of alginate beads, film coating, and cryopreservation, with the proposal of a realistic, standardized quality framework.

From the history and early days of studies, initiated by German researcher Albert Bernhard Frank in 1885, to exactly 140 years of research on different groups of mycorrhizae, this book provides a detailed review of what researchers have achieved up until the advent of molecular techniques. The texts aim to delve deeper into the topic and attract more researchers, thereby contributing to mycorrhizal research across all continents.

The authors express their most sincere thanks to the director of the Universidade Federal do Pampa (UNIPAMPA), campus São Gabriel (Rio Grande do Sul State – Brazil), Dr. Luciana Borba Benetti, for the support of this book; to the Laboratório de Taxonomia de Fungos – LATAF (UNIPAMPA) staff for the technical support. Many thanks.

Jair Putzke
Universidade Federal do Pampa,
São Gabriel, Brazil

Section 1

Introduction

Chapter 1

Introductory Chapter: Why To Study Mycorrhiza

Jair Putzke

1. Introduction

The term "mycorrhiza" was coined by German researcher Albert Bernhard Frank in 1885 when he first recognized a fungus symbiotically interacting with plant roots. The term comes from the Greek words "mykes," meaning "fungus," and "rhiza," meaning "root." The same mycologist first described the term symbiosis in 1877. Arbuscular mycorrhizal (AM – 72% of the total), orchid mycorrhiza (10%), and ectomycorrhiza (ECM – 2%) are the most widely distributed in the plant kingdom, with the ectoendomycorrhiza, ericoid mycorrhiza, monotropoid mycorrhiza, and arbutoid mycorrhiza types being more restricted to a few botanical families [1]. The AM are all from the phyllum Glomeromycota, and the most common genera are *Glomus*, *Gigaspora*, *Scutellospora*, *Acaulospora,* and *Entrophospora*. They are not host-specific and cannot be cultivated under artificial media.

Mycorrhizal relationships involve symbiotic associations between fungi and plants in about 80% of the known taxa, including species of Bryophyta, Pteridophyta, and Gymnosperms and Angiosperms. Many species are highly dependent on this relationship, and fungi can reproduce and disperse independently of the plant species. Many species have not yet been cataloged by science and can reveal more about the complexity of this relationship, which is only possible by attracting and training new scientists for this task. In general, biologists are the main group of scientists studying taxonomy and systematics, but the applications are also worked on by other professionals. A great effort is needed to find the taxa associated with plants before extinction, with taxonomy being the area where more scientists are needed and whose training is only possible through review studies. New fields of study and the application of mycorrhizae are also emerging and need other professionals to assess these fields.

The taxonomy and systematics of this group have gone through different stages since the advent of molecular biology, but surprisingly, it remains an unattractive field for young researchers. The group includes organisms that form macroscopic reproductive structures (like mushrooms and truffles) and others that are exclusively microscopic, being mostly found as spores and hyphae associated with plant roots.

In the obligate symbiotic or mutualistic association, the fungi are found recovering and/or penetrating the roots, receiving carbohydrates (up to 20% of the carbon fixed through photosynthesis) and some other organic compounds from the plant, and exchanging soil nutrients required for plant growth [2].

Microscopic identification, associated primarily with spore characteristics and the obligate biotrophic relationship with the plant, leads to difficulties in morphological identification and, at the same time, in the production of inoculum, which is so necessary for sustainable agriculture [3]. Even though practical commercial applications are already well established, we still have many gaps in terms of application, especially with the use of more species and even in the prospecting of these species, their description, and testing.

In the field of application, mycorrhizae have proven important in contributing to the better development of the associated plant, by bringing nutrients to it, modifying the microbiota associated with the roots, and beneficially altering the characteristics of the soil where they live, being associated with nutrient cycling, carbon sequestration, and soil aggregation [4].

The relationship between these fungi and plants can lead to the development of sustainable agriculture and even to ecosystem services that are not fully understood. They also serve as biostimulants and as biological control agents, reducing biotic and abiotic stress conditions [5]. Ectomycorrhizae can form macroscopic and sometimes very large mushrooms above and underground, which serve as food for the web chain, being an important food resource.

2. Conclusion

New fields have joined the work of understanding the mycorrhiza, especially to elucidate the symbiotic relationship, including metatranscriptomics and metabarcoding, which associate genes with the work that these fungi perform. However, it is still unclear how the fungi benefit from this relationship or how the root microbiome influences the relationship [6, 7]. It is important to prepare more young scientists for these areas in the near future.

Much work with mycorrhizal fungi is still needed before they can be used to truly revolutionize agriculture in the near future, but this depends on more young people embracing the cause and investigating the many questions surrounding this relationship between organisms from two such different kingdoms. To attract the new generation of scientists to these matters of study is essential to guarantee the evolution of this area.

Author details

Jair Putzke
Universidade Federal do Pampa, Bagé, Brazil

*Address all correspondence to: jrputzkebr@yahoo.com

References

[1] W SHI, et al. Diversity and function of mycorrhizal fungi. In: Rigobelo E, editor. Unveiling the Mycorrhizal World. London: IntechOpen; 2024. p. 98

[2] Smith SE, Read DJ. Mycorrhizal Simbiosos. Great Britain: Elsevier Ltd; 2008

[3] SINGH SR, Singh U, Chaubey AK, Shiv R.Mycorrhizal fungi for sustainable agriculture-A review. Agricultural Reviews. 2010;**31**(2):79–92

[4] ZHANG J, Zhao R, LI X, Zhang J. Potential of arbuscular mycorrhizal fungi for soil health: A review. Pedosphere. 2024;**34**(2):279–288

[5] Delaeter M, Magnin-Robert M, Randoux B, Sahraoui AL-H. Arbuscular Mycorrhizal Fungi as Biostimulant and Biocontrol Agents: A Review. Microorganisms. 2024;**12**(7):1281. DOI: 10.3390/microorganisms12071281

[6] Bennett AE, Groten K. The Costs and Benefits of Plant–Arbuscular Mycorrhizal Fungal Interactions. Annual Review of Plant Biology. 2022;**73**:649–672. DOI: 10.1146/annurev-arplant-102820-124504

[7] YOUNG JW. A molecular guide to the taxonomy of arbuscular mycorrhizal fungi. New Phytologist. 2012;**193**:823–826. DOI: 10.1111/j.1469-8137.2011.04029.x

Section 2

Research Advances, Application and Future Perspectives of Mycorrhizal Fungi

Chapter 2

Mycorrhizal Fungi: Research Advances, Applications, and Future Perspectives

Vanielle Galiano Silveira, Ana Luiza Klotz Neves, Laura Amália Miranda Costa, Flavia Helena Aires Sousa, Alice Lemos Costa, Cassiane Furlan Lopes, Jorge Renato Pinheiro Velloso, Marcos André Pinheiro Velloso, Fernando Augusto Bertazzo-Silva and Jair Putzke

Abstract

Mycorrhizae are symbiotic associations between most terrestrial plants and certain groups of fungi, occurring in over 90% of plant species and across nearly all terrestrial ecosystems. In this mutualism, plants provide carbon to fungi, which, in return, supply soil-derived nutrients and confer additional benefits, such as enhanced growth and reduced susceptibility to environmental stresses like drought and salinity. The main types of mycorrhizae are arbuscular mycorrhizae (AM), ectomycorrhizae, ectendomycorrhizae, arbutoid, monotropoid, orchid mycorrhizae, and ericoid mycorrhizae. Advances in the study of the hyphosphere microbiome have clarified the role of fungal hyphae in soil processes and global carbon cycling, with estimates suggesting that nearly one-third of annual fossil fuel CO_2 emissions are directed to mycelial biomass. Mycorrhizal fungi are increasingly studied as bioprotectors, improving crop growth in degraded soils by enhancing water and nutrient uptake, reducing the need for chemical fertilizers, and activating plant defense mechanisms. In the context of climate change, especially with more frequent droughts, AM fungi can help mitigate water stress in crops. Molecular biology tools have revealed thousands of previously undetectable mycorrhizal fungi species, many of which show low host specificity and can be associated with diverse plant species. This has stimulated interest in artificial mycorrhizae, though these lack the resilience of natural systems under abiotic stress and disease. Therefore, they are better suited to complement natural associations or be applied in systems where crops do not naturally form mycorrhizae, contributing to sustainable agriculture and carbon sequestration.

Keywords: symbiosis, ectomycorrhiza, interactions, agriculture, ecosystem

1. Introduction

Throughout evolution, ecological interactions have become increasingly specialized and efficient. Among these, mutualistic relationships stand out, characterized by reciprocal and beneficial exchanges between organisms of different species, which are essential for the functioning of ecosystems to this day [1]. A key example of mutualism occurs between terrestrial plants and fungi associated with their roots, known as mycorrhizae.

Mycorrhizae were first described by [2], and their name derives from the Greek words mykos (fungus) and rhiza (root). These fungi are of great ecological importance, as they contribute to the survival success of host plants by supplying nutrients and providing protection against pathogenic agents. Mycorrhizae can be classified based on morphological characteristics and the plant families with which they associate. According to [3], the main types include arbuscular mycorrhizae (AM), ectomycorrhizae, ectendomycorrhizae, arbutoid, monotropoid, orchid mycorrhizae, and ericoid mycorrhizae.

Mycorrhizal fungi are responsible for absorbing nutrients – particularly phosphorus – and promoting plant growth, in addition to contributing to soil structure and health [4]. These characteristics also make them relevant for economic applications, such as in agriculture. In addition to their ecological importance, mycorrhizal fungi have gained prominence in the economic sphere, as the majority of agricultural crops serve as hosts to arbuscular mycorrhizal fungi, allowing for increased productivity through inoculation with these microorganisms [5].

Despite advances in recent decades, significant gaps remain in the understanding of mycorrhizal functional diversity, their distribution across different ecosystems, and their full potential for application in productive contexts. In this context, it becomes essential to compile existing knowledge, fostering connections between scientific discoveries, practical applications, and future research directions.

Accordingly, this work aims to provide a comprehensive overview of mycorrhizae, exploring their conceptual, morphological, and ecological aspects, as well as their main classifications and functions. Furthermore, it seeks to highlight their practical applications and discuss the challenges and opportunities for research and use of these associations in the contemporary context, especially in the face of ongoing environmental transformations.

2. Definition and classification of mycorrhizal associations

Mycorrhizae occur in more than 90% of plant species and are distributed across virtually all ecosystems [6]. In this interaction, plants supply carbon to fungi, which, in turn, provide soil nutrients to the plants, along with additional non-nutritional benefits such as the stimulation of plant growth and reduced susceptibility to environmental stresses like drought and salinity.

Depending on their function, structure, and the specific plant and fungal species involved [3], mycorrhizae are categorized into four main types: AM, ectomycorrhiza (EM), ericoid mycorrhiza (ErM), and orchid mycorrhiza (OrM). AM is the most ancestral type, having emerged with the earliest land plants [7], in which fungi from the phylum Glomeromycota colonize the roots of approximately 80% of plant species, forming intracellular structures such as arbuscules or coils [3].

On the other hand, the EM association evolved independently multiple times throughout the evolutionary history of plants and fungi [8]. This partnership mainly involves trees that dominate forest and woodland communities in boreal, Mediterranean, and temperate regions, as well as some tropical forests. EM fungi belong to the phyla Basidiomycota, Ascomycota, and, less frequently, Zygomycota, forming a sheath of hyphal filaments around the root tips and penetrating the intercellular space, thereby creating a network between cortical cells known as the Hartig net [3]. For a better example, see the differentiation between AM and ectomycorrhiza (**Figure 1**).

Finally, the most recent types of mycorrhizae in plant evolution are OrM and ErM [7]. In both cases, the fungal partners produce coils of hyphae inside the root cells of orchid and Ericaceae species, respectively. Although the hyphae penetrate the cell wall, the plasma membrane remains intact, allowing for the exchange of substances. While orchid mycorrhizae are found in nearly all species of Orchidaceae, which represent about 10% of vascular plants, ErM occur in some species of Ericaceae, accounting for only 1.5% of all vascular plants.

Despite this relatively limited plant diversity, ErM inhabit a wide range of environments, including Northern Hemisphere heathlands, open forests, Mediterranean woodlands, tropical cloud forests, and sandy lowlands in Australia [9]. Moreover, plants that form ErM colonize ecosystems responsible for more than 20% of the global terrestrial carbon stock, playing a crucial role in the soil carbon cycle.

Nevertheless, ErM remains the least studied and least understood type of mycorrhizal association. More recently, a new subtype of ectomycorrhiza has been described in a restinga area in southern Brazil: the guapirioid ectomycorrhiza, named after the association between the fungus and the plant Guapira opposita (Vell.) Reitz, from the family Nyctaginaceae [10].

In this context, mycorrhizal associations have been fundamental allies that enabled plants to colonize terrestrial environments and continue to sustain the balance of most ecosystems today. From AM, which have been present since the earliest stages of plant life on land, to more specialized forms such as ericoid mycorrhizae (ErM), which are essential in organic matter-rich environments yet remain understudied, and even to recently described types like the guapirioid ectomycorrhiza, all of these interactions reflect a trajectory of coevolution and continuous adaptation.

Understanding this diversity – both morphological and evolutionary – is essential to grasp how these partnerships directly contribute to plant health and development, enhancing nutrient uptake efficiency, increasing tolerance to environmental stress, and maintaining the stability of the ecosystems in which they occur.

3. Fungus–plant symbiosis: Foundations and ecological importance of mycorrhizae

In forest ecosystems, symbiotic interactions play a fundamental role in sustaining life, contributing to the maintenance of moisture, the provision of shelter, and the availability of food for various species of animals, plants, bacteria, and fungi [11].

This symbiosis is particularly advantageous for plants, which are sessile organisms that rely entirely on their surrounding environment to obtain the

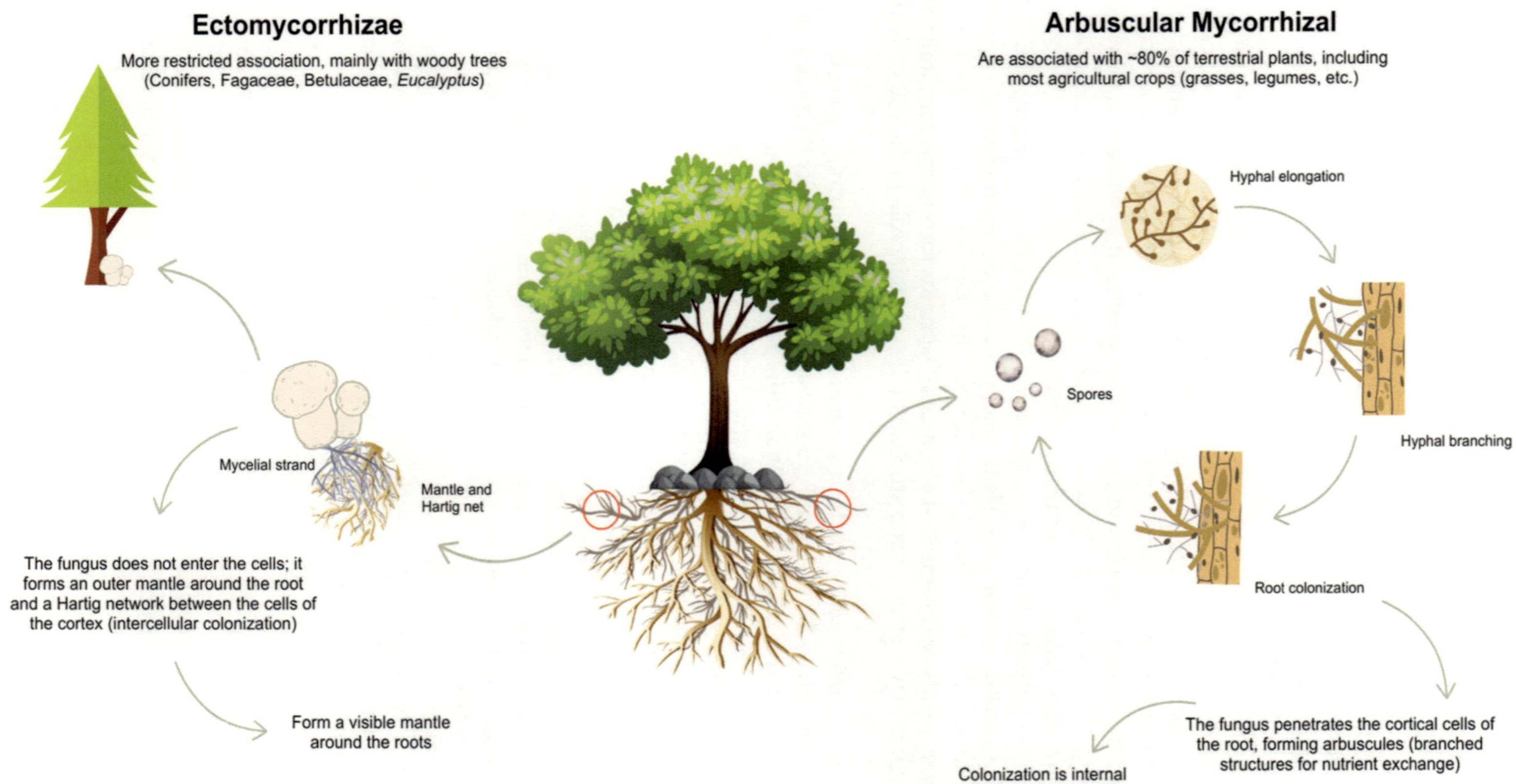

Figure 1.
Diagram illustrating the structural differences between arbuscular mycorrhiza (AM) and ectomycorrhiza. The AM panel shows fungal hyphae penetrating root cortical cells and forming root colonization, whereas the ectomycorrhiza panel shows a fungal mantle surrounding the root and a Hartig net formed between root cortical cells. (source: the authore).

nutrients necessary for their survival. Fungi, on the other hand, are able to expand across large areas of soil through their hyphae, reaching nutrients in regions beyond the reach of plant roots. In this way, they act as extensions of the root system, promoting the efficient absorption of water and minerals [12].

The mycorrhizal association is essential for plant vitality. By acquiring nutrients provided by fungi, plants are able to carry out their physiological and ecological functions, including photosynthesis, which produces glucose as a final product. A portion of this glucose and carbon is then transferred to the fungi, completing the exchange cycle and sustaining the symbiosis [13]. This is, therefore, a highly efficient mutualistic interaction that is beneficial to both organisms.

In addition to the exchange of nutrients and energy, mycorrhizae also promote other essential functions, such as enhancing plant tolerance to various types of stress. This includes both abiotic stresses – such as drought, salinity, and temperature fluctuations – and biotic stresses, such as attacks by pests and pathogens. In this way, mycorrhizae contribute significantly to the adaptation and survival of plants in challenging environments [14].

By integrating knowledge of mycorrhizal interactions with recent scientific advances, it becomes evident that these associations extend beyond the fungus–plant nutritional exchange. A complex, multifunctional ecological network connects a wide diversity of organisms, and further insight into the connections within the mycorrhizosphere may enhance the sustainable applicability of these associations.

Thus, given their ecological, functional, and evolutionary importance, mycorrhizae not only support the health and productivity of ecosystems but also play a strategic role in environmental resilience in the face of global change. Nevertheless, despite their recognition as key elements in ecological dynamics, many aspects of their biology, ecology, and interactions remain poorly understood. In this context, recent advances in mycorrhizal fungi research have been crucial for elucidating previously unknown processes – from the structure of the mycorrhizosphere to the role of the microbiome associated with fungal hyphae.

4. Practical applications of mycorrhizae in agriculture

Mycorrhizal associations are found in both conventional and functional systems, particularly in areas where carbon sequestration and soil aggregation are most prevalent. These associations occur in conjunction with other organisms in the community, within which the fungal mycelium develops a specific ecological niche. Associations with other organisms are diverse in multiple ways. Soil health is one of the most critical factors, and various associations with microorganisms, plants, and animals have been identified. Through this network, a functional group emerges that contributes to multiple ecological processes [15].

In recent years, advances in research on the hyphosphere microbiome have elucidated the functional roles of hyphae and their contributions to key soil processes, while also highlighting the potential for harnessing AMF and AMF-associated microbiota. AM have undergone shifts in their ecological functions, particularly in relation to root symbioses [16].

A recent study [17] highlights important points involving AM, emphasizing that most vascular plants are mycorrhizal, their families can typically be mycorrhizal or not, colonization generates plant benefits, and the ecology and distribution of mycorrhizal fungi depend on the mycorrhizal type. In contrast, some concepts are new and are still being studied. One of these is the core microbiome associated with the hyphae of AM fungi. The core microbiome is a crucial aspect that should be further explored using more effective methods. This may involve the transformation of nutrients, such as nitrogen (N), through mineralization to meet fungal requirements [18].

The relationships between plants and mycorrhizal fungi become effective in the presence of available nutrients, and this interaction results in a bridge between plants, soils, and humans. Carbon nutrients are essential for this bridge to occur. Mycorrhizal fungi are important in practical applications such as C cycling, where over 36% of the CO2 annual emissions from fossil fuels are allocated to the mycelia [19]. The degradation of some nutrients into mycelial fungi has the potential to engineer non-legume plants and is one of the most explored points in recent analyses. The description of the cytological changes and the genes involved has been analyzed; however, the structure and function of the mycorrhizosphere are still largely unexplored [20].

For advances utilizing mycorrhizal fungi, plant seeds have accelerated and restored post-mining landscapes [21]. Enzymatic activity and the presence of functional genes were described as increasing with improved absorption efficiency through the presence of mycorrhizal fungi and bacteria from the soil core. Various analyses have been performed using next-generation sequencing (NGS). This method revealed an interaction between the beneficial microbiome and the mycorrhizal fungi, elucidating alterations in root exudation and generating better nutrient availability for the microbiome to support plant growth [22].

Mycorrhizal fungi have enormous potential for use in several areas. Although they possess widely known features, research on this group still presents several critical questions. One of these is the understanding of their ecological functions. Through studies such as the expression of certain genes and what occurs after the inoculation process involving their genetic role, some researchers have been able to explore aspects of the carbon sink and encompass other areas [23]. Many recent studies have evaluated mycorrhizal fungi as bioprotectors of crops against a pathogen of the genus *Verticillium* and highlighted that interference from climate change may cause deterioration of extraradical mycelia [24].

Analyses against *Verticillium* were also performed using compost to improve the growth and yield of tomato and its resistance to *Verticillium dahliae*. As a result, mycorrhized and compost-amended soils most effectively reduced plant malondialdehyde accumulation when *V. dahliae* was present [25]. Alternatively, mycorrhizal fungi induce the activation of the tomato defense system, mitigating *Fusarium* wilt stress and increasing antioxidant enzyme levels [26].

In another context, mycorrhizal fungi have been utilized to evaluate their interference with the growth of endemic bunchgrass in pine savannas, elucidating that the fungi may improve plant interactions in disturbed soils by providing essential nutrients [27]. Many advances have been made in the study of mycorrhizal fungi (**Figure 2**), and with the advancement of certain analyses, we can better understand how they function and how their relationships are maintained. In addition, the association between fungi and plants is essential for our ecosystems.

In this way, we can enjoy the different benefits provided by this correlation in several areas, including the improvement of crops and food quality, as well as indicators of environmental and soil quality, among others. Some aspects need to be further explored and studied because, due to the large number of factors involved in mycorrhizal

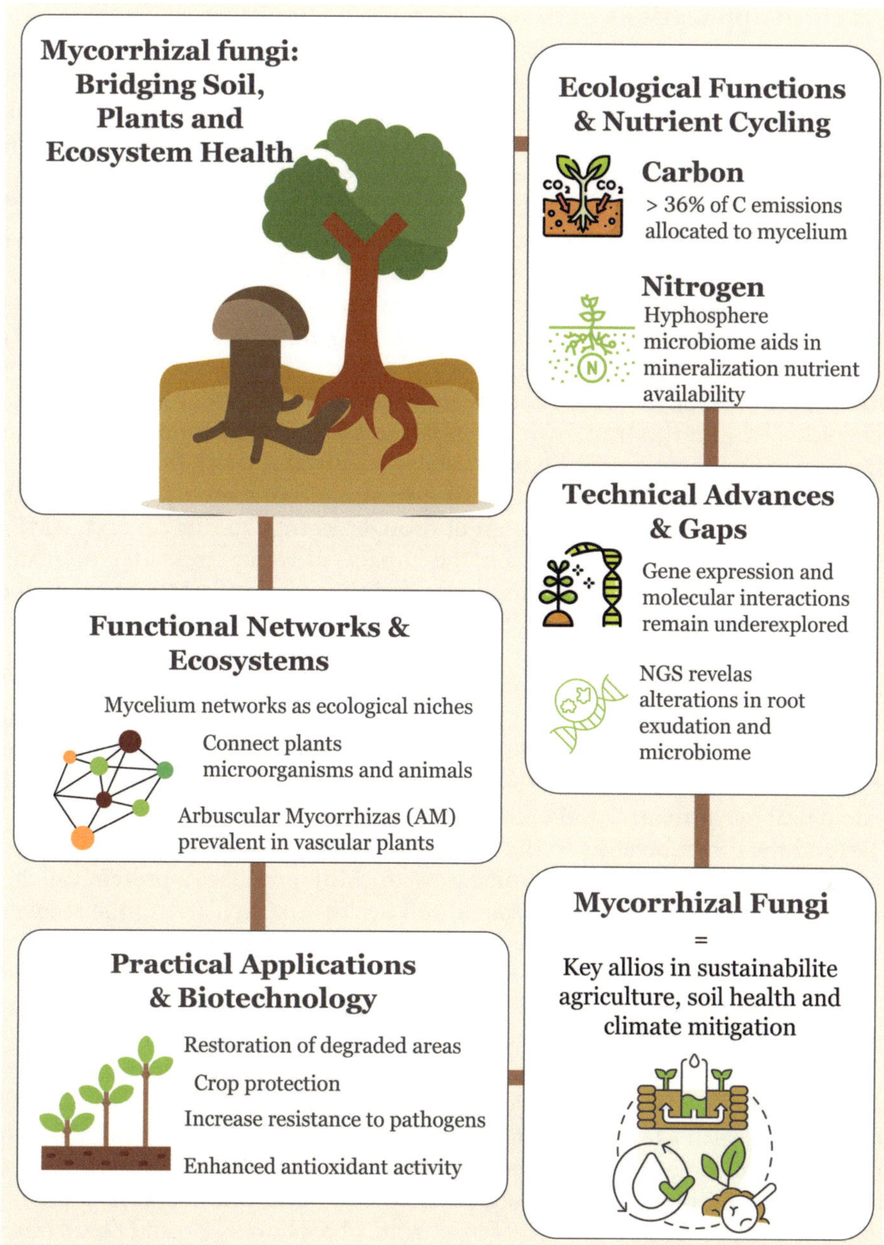

Figure 2.
Schematic overview of recent advances in mycorrhizal research and applications, highlighting progress in molecular and genomic tools, improved understanding of mycorrhizal ecology and plant–fungus interactions, and practical applications in agriculture, forestry, ecosystem restoration, and sustainable soil management (source: the authors).

associations, comprehensive analyses are extremely necessary. There is a growing need to enhance agricultural productivity, resilience, and soil restoration. Therefore, future perspectives must sustainably integrate both practical applications and scientific research, as tangible benefits are observed in the field while studies continue to uncover the complexity of symbiotic networks.

5. Practical applications of mycorrhizae in agriculture

Mycorrhizal symbioses have gained prominence in agricultural and environmental sciences not only for their ecological role but also for their multiple applications in more sustainable production systems. In recent decades, advances in understanding the molecular mechanisms of this symbiosis, combined with the development of commercial inoculants and efficient management strategies, have made it possible to apply mycorrhizal fungi on a large scale – meeting the growing demand in this field for ecological solutions in the agricultural sector [28].

Among the main groups, arbuscular mycorrhizal fungi (AMF), belonging to the phylum *Glomeromycota*, have emerged as promising allies in sustainable agriculture. Their association with plant roots significantly enhances the uptake of water and nutrients, reduces the need for chemical fertilizers, and provides plant defense mechanisms [29]. In this way, AMF act as natural biofertilizers and valuable allies in addressing the impacts caused by intensive agricultural activity [30].

One of the major current challenges to agricultural productivity concerns climate change, particularly the intensification of drought events. In this context, AMF serves as important allies in mitigating the impacts of water stress on crop plants. A study conducted by LI et al. [31] investigated the effects of AMF inoculation on maize (*Zea mays*) crops. According to the authors, the inoculation significantly improved drought resistance, as reflected by increased plant biomass, greater leaf water retention, and activation of antioxidant enzymes. Complementarily, Ouhaddou et al. [32] evaluated the effects of dual inoculation – with AMF and plant growth-promoting rhizobacteria (PGPR) – also in maize. The results demonstrated that this combination enhanced tolerance after inoculation, promoted improved physiological performance, and enhanced soil health.

Beyond the direct benefits to the plant, AMF positively affects soil structure and quality. Chemically, during hyphal growth, AMF produces a protein called glomalin. This protein plays a key role in soil health – particularly in the recovery of degraded soils – as it functions as a "glue" that stabilizes aggregates and contributes to carbon and nutrient retention. Glomalin is present in the hyphal walls and spores, accumulating in the soil over time. This makes it especially important in the restoration of degraded areas, contributing to overall soil health and fertility [33].

The application of AMF has also demonstrated potential for promoting the biofortification of food crops, particularly regarding the uptake of essential micronutrients such as iron (Fe) and zinc (Zn). Founoune-Mboup et al. [34] evaluated that inoculation with *Glomus mosseae* in pearl millet (*Pennisetum glaucum*), in combination with organic residues, resulted in a significant increase in plant biomass and higher concentrations of iron, zinc, phosphorus (P), and potassium (K) in the aerial parts of the plants compared to non-mycorrhizal controls.

Nguyen et al. [35] investigated the effects of *Rhizophagus irregularis* inoculation in rice plants grown under varying phosphorus levels and aerobic conditions. Despite the overall reduction in Zn bioavailability, results showed that the fungi enhanced Zn and Fe concentrations in the aleurone layer of the grains. Another strategic role of AMF lies in their use as phytoremediators of soils contaminated by heavy metals. Their association with plant roots facilitates the immobilization of metals in the root system, a process known as phytostabilization [36].

AMF acts as physical and chemical barriers, forming a protective sheath around roots that blocks xenobiotics through its mycelium. Moreover, AMF has been observed in contaminated environments, demonstrating potential to support plant growth under metal-induced stress. This strategy is considered promising due to its ecological sustainability, cost-effectiveness, and low generation of secondary waste [36].

AMF are essential for the development of more efficient, productive, and sustainable agricultural practices. Their multifunctional role ranges from improving plant nutrition to the restoration of degraded and contaminated soils, as well as mitigating abiotic stresses and promoting food biofortification. When combined with scientific advancements, this practice enhances the potential and applicability of symbiosis as an agro-environmental tool. Scientific progress serves as a strategic foundation, as understanding processes such as carbon sequestration, nutrient cycling, and protective actions against pathogens, for instance, enables the integration of theoretical knowledge into practical methodologies applicable in the field.

6. Future perspectives for the use of mycorrhizae

Mycorrhizal fungi have been employed across a wide range of fields, from seed improvement and increased productivity to enhancing plant resilience to climate change and even more unexpected applications, such as the bioremediation of degraded areas. One of the most frequently addressed topics – and the focus of a substantial portion of the research – is the relationship between plants and AMF. In a systematic review, Bennett and Groten [14] point out that the costs and benefits of the AM–plant symbiosis have mostly been characterized through laboratory response experiments, while the mechanisms underlying this relationship remain poorly understood in natural environments.

Yet, how can we quantify the benefits of these interactions for mycorrhizal fungi, considering they cannot be cultured independently? Among the known advantages of these fungi, the primary ones include access to carbon – primarily in the form of lipids – as well as water and shelter. For the associated plants, the benefits include improved uptake of nitrogen and other soil nutrients, resulting from the increased root surface area [37]. Moreover, with the advent of molecular biology, it has become possible to detect thousands of mycorrhizal fungal species previously undetected, many of which exhibit low host specificity and can associate with a wide variety of plants [38].

Based on these discoveries and the advancement of genetic engineering techniques, research into the synthesis of artificial mycorrhizae is expanding. However, it is already known that such artificial mycorrhizae lack the same resilience to abiotic stress and disease. They should, therefore, be used to complement natural systems or in agricultural settings involving cultivars that do not naturally form these associations – thereby reducing the need for fertilizers and contributing to carbon sequestration [39].

In order to maximize the potential of mycorrhizae in scientific development, Mujica et al. [40] recommend improvements aimed at developing and integrating global databases on mycorrhizae. They highlight the importance of sharing both published and unpublished data, as well as standardizing ecological testing to enable the replication of results. In this way, the integration and globalization of data concerning the behavior, identity, interactions, and distribution of mycorrhizal species, starting from the regional scale, will increasingly contribute to our understanding of these organisms.

7. Conclusion(s)

Mycorrhiza plays a crucial role in maintaining the ecological balance of ecosystems. Through this symbiotic association, many species are able to survive and thrive even under adverse environmental conditions. The mutualistic exchange between organisms from different kingdoms – particularly between the Fungi and Plantae kingdoms – highlights the efficiency of symbiosis as an adaptive strategy and a fundamental mechanism for sustaining life within ecosystems.

Beyond their fundamental ecological roles, the applicability of mycorrhizae has attracted significant scientific interest and investment in research. Recent studies have explored the potential of these associations across various fields, ranging from sustainable agriculture – through improved nutrient uptake and reduced fertilizer use, as previously discussed – to applications in biotechnology, reforestation, restoration of degraded areas, and the mitigation of climate change effects.

Thus, mycorrhizae not only support natural ecosystems but also represent a valuable tool for promoting more balanced, resilient, and efficient environmental practices. It is important to note, however, that there remains considerable scope for further research to deepen our understanding of their ecological and economic potential, contributing to more sustainable solutions for addressing contemporary environmental challenges.

Acknowledgments

We would like to thank the team of Laboratório de Taxonomia de Fungos (LATAF) of Universidade Federal do Pampa (UNIPAMPA) for the support provided to this research. This study was financed in part by the Coordenação de Aperfeiçoamento de Pessoal de Nível Superior – Brasil (CAPES) – Finance Code 001, and by Fundação de Amparo à Pesquisa do Estado do Rio Grande do Sul (FAPERGS) under nbr. 21/2551-0001985-9 and Conselho Nacional de Pesquisa (CNPQ) nbr. 405564/2022-8.

Conflict of interest

The authors declare no conflict of interest.

Mycorrhizal Fungi: Research Advances, Applications, and Future Perspectives
DOI: http://dx.doi.org/10.5772/intechopen.1013590

Author details

Vanielle Galiano Silveira*, Ana Luiza Klotz Neves, Laura Amália Miranda Costa, Flavia Helena Aires Sousa, Alice Lemos Costa, Cassiane Furlan Lopes, Jorge Renato Pinheiro Velloso, Marcos André Pinheiro Velloso, Fernando Augusto Bertazzo-Silva and Jair Putzke*
Laboratório de Taxonomia de Fungos, Universidade Federal do Pampa, São Gabriel, Rio Grande do Sul, Brazil

*Address all correspondence to: vaniellegaliano@gmail.com and jrputzkebr@yahoo.com

References

[1] Bronstein JL. Estudando Mutualismo. em Mutualismo. Bronstein JL, editor. Oxford: Oxford University Press, 2015; p. 4–19

[2] Frank B. Über die auf Wurzelsymbiose beruhende Ernährung gewisser Bäume durch unterirdische Pilze. Berichte der Deutschen Botanischen Gesellschaft. 1885;**3**:128–145

[3] Smith SE, Read D. Mycorrhizal Symbiosis. 3th ed. New York, EUA: Academic Press; 2008. p. 780

[4] Schaefer DA, Gui H, Mortimer PE, Xu J. Micorriza arbuscular e agricultura sustentável. Sistemas Agrícolas Circulares. 2021;**1**(1):1–7

[5] Chen M, Arato M, Borghi L, Nouri E, Reinhardt D. Serviços benéficos de fungos micorrízicos arbusculares – da ecologia à aplicação. Frontiers in Plant Science. 2018;**9**:1270

[6] Van Der Heijden MGA, Martin FM, Selosse MA, Sanders IR. Mycorrhizal ecology and evolution: The past, the present, and the future. New Phytologist. 2015;**205**:1406–1423

[7] Strullu-Derrien C, Selosse MA, Kenrick P, Martin FM. The origin and evolution of mycorrhizal symbioses: From palaeomycology to phylogenomics. New Phytologist. 2018;**220**:1012–1030

[8] Tedersoo L, Brundrett M. Chapter 19: Evolution of ectomycorrhizal symbiosis in plants. In: Tedersoo L, editor. Biogeography of Mycorrhizal Symbiosis, Ecological Studies 230. Switzerland: Springer; 2017. p. 407–467

[9] Cairney JWG, Meharg AA. Ericoid mycorrhiza: A partnership that exploits harsh edaphic conditions. European Journal of Soil Science. 2003;**54**:735–740

[10] Furtado AN, Leonardi M, Comandini O, Rinaldi AC, Neves MA. Guapirioid ectomycorrhiza: A novel fungus-plant subtype is described associated to Guapira opposita (Nyctaginaceae) in the Brazilian restinga. Forest Systems. 2023;**32**(2): e009–e009

[11] Deveau A, et al. Fungos Fitopatogênicos de Solo em Sistemas Agroflorestais: revisão de literatura. FEMS Microbiology Reviews 2018;**42**:335–352

[12] Agápto JP, et al. Fungos Fitopatogênicos de Solo em Sistemas Agroflorestais: revisão de literatura. Diversitas Journal 2021;**6**:3052–3079

[13] Ganiko-dutra M, DE Lara Weiser V.Quem são os fungos e por que precisamos conservá-los? Aprendendo Ciência (ISSN 2237-8766). 2019;**8** (1):42–47

[14] Bennett AE, Groten K. The costs and benefits of plant–arbuscular mycorrhizal fungal interactions. Annual Review of Plant Biology. 2022;**73** (1):649–672

[15] Shen RF, Teng Y. The frontier of soil science: Soil health. Pedosphere. 2023;**33**:6–7. DOI: 10.1016/j. pedsph.2022.06.007

[16] LI CJ, Li HG, Hoffland E, Zhang FS, Zhang JL, Kuyper TW. Common mycorrhizal networks asymmetrically improve chickpea N and

P acquisition and cause overyielding by a millet/chickpea mixture. Plant and Soil. 2022b;**472**:279–293. DOI: 10.1007/s11104-021-05232-0

[17] Albornoz FE, Dixon KW, Lambers H. Revisiting mycorrhizal dogmas: Are mycorrhizas really functioning as they are widely believed to do? Soil Ecology Letters. 2021;**3**:73–82. DOI: 10.1007/s42832-020-0070-2

[18] Wang LT, George TS, Feng G. Concepts and consequences of the hyphosphere core microbiome for arbuscular mycorrhizal fungal fitness and function. New Phytologist. 2023. DOI: 10.1111/nph.19396

[19] Hawkins HJ, Cargill RIM, Van Nuland ME, Hagen SC, Field KJ, Sheldrake M, Soudzilovskaia NA, Kiers ET. Mycorrhizal mycelium as a global carbon pool. Current Biology. 2023;**33**:R560–R573. DOI: 10.1016/j.cub.2023.02.027.

[20] SHI JC, Wang XL, Wang E. Mycorrhizal symbiosis in plant growth and stress adaptation: From genes to ecosystems. Annual Review of Plant Biology. 2023;**74**:569–607. DOI: 10.1146/annurev-arplant-061722-090342

[21] Vahter T, Bueno CG, Davison J, Herodes K, Hiiesalu I, Kasari-Toussaint L. Co-introduction of native mycorrhizal fungi and plant seeds accelerates restoration of post-mining landscapes. Journal of Applied Ecology. 2020;**57**:1741–1751. DOI: 10.1111/1365-2664.13663

[22] XU YJ, Chen Z, Li XY, Tan J, Liu F, Wu JP. Mycorrhizal fungi alter root exudation to cultivate a beneficial microbiome for plant growth. Functional Ecology. 2023;**37**:664–675. DOI: 10.1111/1365-2435.14249

[23] Xue H, Wang W, Wang X, et al. The potential of mycorrhizal fungi to increase terrestrial ecosystem carbon sink: A review. Eurasian Soil Science. 2023;**56**:1724–1738

[24] Goicoechea N. Mycorrhizal fungi as bioprotectors of crops against verticillium wilt—A hypothetical scenario under changing environmental conditions. Plants. 2020;**9**(11):1468. DOI: 10.3390/plants9111468

[25] Ait Rahou Y, Ait-El-Mokhtar M, Anli M, Boutasknit A, Ben-Laouane R, Douira A, ... Meddich A. Use of mycorrhizal fungi and compost for improving the growth and yield of tomato and its resistance to Verticillium dahliae. Archives of Phytopathology and Plant Protection. 2021;**54**(13–14):665–690. DOI: 10.1080/03235408.2020.1854938

[26] Hashem A, Akhter A, Alqarawi AA, Singh G, Almutairi KF, Abd_Allah EF. Mycorrhizal fungi induced activation of tomato defense system mitigates Fusarium wilt stress. Saudi Journal of Biological Sciences. 2021;**28**(10):5442–5450. DOI: 10.1016/j.sjbs.2021.07.025

[27] Kimball A, Love D, Lopez-Scarim V, Zee A, Crandall RM, Karlsen-Ayala E, Fill JM. Mycorrhizal fungi affect growth of an endemic bunchgrass in pine savannas. Ecological Solutions and Evidence. 2025;**6**(2):e70026. DOI: 10.1002/2688-8319.70026

[28] Ho-plágaro T, García-Garrido JM. Molecular regulation of arbuscular mycorrhizal symbiosis. International Journal of Molecular Sciences. 2022;**23**(11):5960. DOI: 10.3390/ijms23115960

[29] WANG Q, Liu M, Wang Z, Li J, Liu K, Huang D. The role of arbuscular mycorrhizal symbiosis in plant abiotic stress. Frontiers in Microbiology. 2024;**14**. DOI: 10.3389/fmicb.2023.1323881.

[30] Cavagnaro TR, Bender SF, Asghari HR, Van Der Heijden MG. The role of arbuscular mycorrhizas in reducing soil nutrient loss. Trends in Plant Science. 2015;**20**:283–290. DOI: 10.1016/j.tplants.2015.03.004

[31] LI J, Zhou L, Chen G, Yao M, Liu Z, Li X, . . . Chen X. Arbuscular mycorrhizal fungi enhance drought resistance and alter microbial communities in maize rhizosphere soil. Environmental Technology & Innovation. 2025;**37**:103947. DOI: 10.1016/j.eti.2024.103947

[32] Ouhaddou R, Ech-Chatir L, Ikan C, Soussani FE, Errouh F, Boutasknit A, . . . Meddich A. Investigation of the impact of dual inoculations of arbuscular mycorrhizal fungi and plant growth-promoting rhizobacteria on drought tolerance of maize grown in a compost-amended field under Mediterranean conditions. Frontiers in Microbiology. 2024;**15**. 1432637. DOI: 10.3389/fmicb.2024.1432637

[33] Staunton S, Saby NPA, Arrouays D, Quiquampoix H. Can soil properties and land use explain glomalin-related soil protein (GRSP) accumulation? A nationwide survey in France. Catena. 2020; **193**:104620. DOI: 10.1016/j.catena.2020.104620

[34] Founoune-mboup H, Diallo B, Adigoun RFR, Kane A, Fall AF. Contribution of arbuscular mycorrhizal fungi to the bioavailability of micronutrients (iron and zinc) in millet accessions. Frontiers in Plant Science. 2024;**15**:1364469. DOI: 10.3389/fpls.2024.1364469

[35] Nguyen TD, Sai N, Johnson AA, Lombi E, Smith E, Doolette CL, Watts-Williams SJ. Arbuscular mycorrhizal fungi increase aleurone layer zinc concentration but reduce overall zinc bioavailability in rice grain. Plants, People, Planet. 2025. DOI: 10.1002/ppp3.10633

[36] Morar F, Iantovics LB, Gligor A. Analysis of phytoremediation potential of crop plants in industrial heavy metal contaminated soil in the Upper Mures River Basin. Journal of Environmental Informatics. 2018;**31**(1). DOI: 10.3808/jei.201700383

[37] Trappe J. A.B. Frank and mycorrhizae: The challenge to evolutionary and ecologic theory. Mycorrhiza. 2005;**15**:277–281

[38] Antoine S, Hériché M, Boussageon R, Noceto PA, van Tuinen D, Wipf D, Courty PE. A historical perspective on mycorrhizal mutualism emphasizing arbuscular mycorrhizas and their emerging challenges. Mycorrhiza. 2021;**31**:1–17

[39] Hornstein ED, Sederoff H. Back to the future: Re-engineering the evolutionarily lost arbuscular mycorrhiza host trait to improve climate resilience for agriculture. Critical Reviews in Plant Sciences. 2024;**43**(1):1–33

[40] Mujica MI, Silva-Flores P, Bueno CG, Duchicela J. Integrating perspectives in developing mycorrhizal trait databases: A call for inclusive and collaborative continental efforts. New Phytologist. 2024;**242**(4):1436–1440

Chapter 3

Progress of Research on the Effects of Arbuscular Mycorrhizal Fungi on Plant–Soil Systems and Application of Mycorrhizal Fungi

Ying Wu Shi, Hongmei Yang, Ming Chu, XinXiang Niu, Huifang Bao, Ning Wang, Faqiang Zhan, Xuanqi Long, Rong Yang, Yun Chen, Shuang Dou, Zhichao Meng, Zhao Zhang and Kai Lou

Abstract

This paper reviewed the research progress and application of arbuscular mycorrhizal fungi (AMF) in the plant–soil system. AMF significantly improves plant growth and soil health by promoting nutrient uptake, enhancing stress resistance, and improving soil structure. In agriculture, AMF helps to reduce the use of chemical fertilizers and pesticides and improve crop yield and quality; in ecological restoration, AMF promotes vegetation restoration of degraded land. Future research should focus on the interaction mechanism between AMF and plants and the optimization of its application technology, so as to give full play to its potential in sustainable agriculture and ecological environment protection.

Keywords: arbuscular mycorrhizal fungi (AMF), plant–soil system, nutrient absorption, resilience, ecological restoration

1. Introduction

Mycorrhiza is a kind of mutualistic symbiont formed by soil fungi and plant roots [1]. Arbuscular mycorrhiza (AM) is one of the most widely distributed mycorrhizal types. AM is formed by the symbiotic development of plant roots and Zygomycota. Arbuscular mycorrhizal fungi (AMF or AM fungi) are important soil fungi that can form AM with more than 80of angiosperms in nature.

AMF is a soil microorganism that is widely distributed and closely associated with plants. AMF exists in many habitats of ecosystems, such as farmland [2], orchards [3], pastoral areas [4], forests [5], shrubs [6], and grasslands [7]. Most terrestrial plants can form a mutually beneficial symbiotic relationship with AMF [8]. Plants

IntechOpen

provide carbon sources for the growth and development of mycorrhizal fungi, and mycorrhizal fungi can help plants better absorb nutrients in the soil, especially P [9]. AMF has many positive effects on plants, such as increasing plant uptake of N and P in soil and promoting plant growth [10]; AMF can promote the growth and development of plants at different stages by increasing stomatal conductance, accumulating soluble sugar, increasing soil stability, improving crop tolerance to various stresses, and maintaining the function of agricultural systems [11].

2. Diversity of mycorrhizal fungi

In the context of global climate change, plant invasion has caused changes such as reduced biodiversity, which has profoundly affected the structure and function of ecosystems [12, 13]. Mycorrhiza, as a symbiont of fungi and plant roots in soil, can form a complex mycelial network in plant roots, improve the absorption and utilization of soil mineral nutrients (such as nitrogen, phosphorus, and sulfur) by host plants, and help host plants adapt to various stress conditions [14, 15]. There are many kinds of mycorrhizal fungi, which can form a good symbiotic relationship with plant roots, improve root architecture (**Figure 1**), enhance antioxidant capacity, and improve plant growth by promoting root nutrient absorption [16].

The agricultural ecosystem is closely related to human life and provides the material basis and guarantee for human survival and development. It is reflected in all aspects of clothing, food, housing, and transportation required for human survival and development. The destruction of the agricultural ecosystem will lead to the consumption of a large number of natural resources and the destruction of the ecological environment, break the normal material circulation and energy flow of the agricultural ecosystem, affect the safety of food and the exertion of agricultural productivity, limit the long-term development of agriculture, and ultimately threaten the development of human beings [17]. AMF directly or indirectly affects soil properties and plant growth and development, so the study of AMF species diversity in agricultural ecosystems is of far-reaching significance.

Figure 1.
The functional versatility of AMF.

2.1 AMF species diversity in terrestrial agroecosystems

AMF belongs to Glomeromycota in the classification system. According to the latest classification system of AMF, there are one phylum, one class, four orders, 11 families, 27 genera, and about 300 species [18]. According to the review of published literature, as of 2018, there are 20 genera and 125 species of AMF found in the soil of terrestrial agricultural ecosystems in China, including 28 species of *Acaulospora* and four species of *Ambispora*. There are two species of Archaeospora, four species of Claroideoglomus, one species of *Corymbiglomus*, two species of Dentiscutata, three species of Diversispora, two species of Dominikia, one species of Entrophospora, and seven species of Funneliformis. There are five species of Gigaspora, 32 species of Glomus, three species of Pacispora, two species of *Paraglomus*, four species of Racocetra, one species of Redeckera, seven species of Rhizophagus, five species of Sclerocystis, and 10 species of Scutellospora. There are two species of Septoglomus. Glomus and Acaulospora have the most species, which are the dominant genera of AMF in terrestrial agricultural ecosystems in China.

2.2 Spatial and temporal distribution characteristics of AMF diversity

The biological community has different habitat selections, and the AMF community also exhibits distinct spatial and temporal distribution characteristics. AMF demonstrates varying species diversity across different spatial and temporal structures.

2.2.1 Spatial structure

In recent years, researchers have conducted extensive studies on the diversity changes in AMF communities across geographical distributions, particularly along altitude gradients. Regarding the trend of microbial diversity changes along altitude gradients, most studies have found that the diversity of AMF communities decreases as altitude increases. For example, Turrini et al. [19] examined the diversity and community structure of AMF in 21 apple orchards located at different altitudes and discovered that low-altitude apple orchards contained more operational taxonomic units (OTUs) than high-altitude apple orchards. The number of AMF species identified in high-altitude apple orchards was lower than that in low-altitude apple orchards, indicating that the AMF species richness in apple tree roots at varying altitudes was significantly different, although there was no significant difference in the community structure of AMF. The results demonstrated that the diversity of AMF decreased with increasing altitude, and the species richness of AMF also declined as altitude increased. The study further revealed that the AMF spore density in low-altitude planting areas was the lowest, the species composition was the most abundant, and the root infection rate was the lowest [20]. These findings indicated that the altitude of the study area was negatively correlated with AMF richness but positively correlated with AMF infection rate. However, it has also been observed that the diversity of AMF species in farmland soils increases with rising altitude [21].

The AMF community is seriously affected by altitude, which is due to the change in temperature and humidity caused by the existence of an altitude gradient. These environmental changes affect the distribution of the microbial community. There is no specific conclusion on the changes in AMF species richness in agricultural ecosystems with latitude changes. Mao et al. [20] found that the latitude of Kunming, Yongren, and Lijiang ranged from high to low by studying the number, infection, and community structure of AMF spores in three olive-planting areas in Yunnan, but the root infection rate, spore density, and species composition of AMF in the three planting areas were not correlated with latitude. Zhou et al. [22] studied six sea buckthorn plots from southwest to northeast China, and the analysis showed that there was no significant difference in the distribution of the Shannon diversity index and OTU richness of AMF at different latitudes. The study by Fierer et al. also found that there was no significant correlation between AMF diversity and the latitude gradient [23]. At present, most of the research on the elevation gradient of AMF is concentrated in the alpine plateau, and the research on latitude is a large-scale study. Therefore, the study of the spatial structure of farmland is a direction for future research work.

2.2.2 Time structure

Most of the studies on the changes in AMF diversity and community composition over timescales are long-term field tillage monitoring studies and time succession. The diversity of AMF is affected by seasons. By studying the diversity of soil AMF communities in sugarcane planting areas during different seasons, it has been found that the community composition in spring, summer, autumn, and winter is quite different. The diversity of AMF in summer and autumn is significantly higher than that in spring and winter. The study by Hu et al. also shows that the diversity of AMF is affected by seasons [24]. In autumn, the relative abundance of AMF species gradually increases. The results of Schultz et al. [25] showed that the sporulation and distribution of AMF exhibit seasonal characteristics. Some AMF species germinate in late spring, while others germinate in late summer. These studies are of great significance for understanding the seasonal community changes of AMF. The succession of ecosystems over time also alters the fungal community. Zhang et al. studied the natural succession of abandoned farmland [26]. After 30 years of succession from abandoned farmland to grassland, it was found that the diversity of fungi increased following natural succession, as observed by studying the succession of abandoned farmland every 5 years from 0 to 30 years. This indicates that agricultural tillage may affect the diversity of AMF in cultivated land, and the community composition of AMF changes after years of natural succession.

2.2.3 Influencing factors of AMF diversity

The species diversity of AMF is affected by many factors, which are regulated by both biological and abiotic factors. In the agricultural ecosystem, human disturbance, specifically agricultural management measures, has a significant impact. Many factors can lead to changes in AMF richness, including climate, vegetation, soil physical and chemical properties, and the interactions between these factors. Chaudhary et al. proposed a conceptual model of AMF biogeography [27] and identified three categories of factors that regulate the distribution of AMF species: abiotic external factors (such as temperature and precipitation), biotic external factors (such as plant communities and interspecific interactions), and intrinsic characteristics of AMF (such as dispersal ability).

2.2.3.1 Agricultural management measures

In agroecosystems, agricultural practices are a major and inevitable factor affecting the distribution of AMF. Agricultural measures include tillage practices, such as tillage, rotation, intercropping, and no-tillage, as well as fertilization measures, such as N and P fertilizer application and spraying pesticides. There is considerable evidence that tillage practices can cause mechanical damage to the AMF mycelial network in the soil and disturb the AMF community, thereby reducing the spore density of AMF and the infection rate of host plants [28]. Jie et al. studied the effects of different continuous cropping years on the diversity of AMF in the rhizosphere soil of soybean at the branching stage [29]. The results of morphological and denaturing gradient gel electrophoresis analysis showed that the increase in continuous cropping years could increase the infection rate of AMF in soybean roots and enhance the richness of the AMF community, which had a significant effect on AMF. This shows that continuous cropping is beneficial for improving the diversity index of the AMF community. In the winter crop–soybean rotation system, the crop yield will gradually decrease with the rotation time, which is due to the continuous rotation causing a decline in AMF growth and infection, as well as nutrient loss in the soil. Compared with traditional tillage methods, no-tillage and permanent planting conservation tillage methods can significantly improve the AMF infection ability of plants. Compared with the traditional maize–wheat rotation system, no-tillage increased the length of AMF hyphae outside the roots [30].

The combined use of no-tillage and mulching crops also significantly increased AMF spore density and species richness. Gaoping et al. analyzed the effects of different tillage measures on AMF diversity in the Loess Plateau [31]. Through the study of four tillage measures (tillage, no-tillage, tillage and straw mulching, no-tillage and straw mulching), it was found that different measures did have an impact on AMF diversity, and AMF diversity decreased in turn under no-tillage, no-tillage and straw mulching, tillage and straw mulching, and tillage methods. This study also verified the positive effect of no-tillage on AMF diversity. Verzeaux et al. [32] reviewed the effects of different agricultural measures on AMF communities, which in turn affected nitrogen-use efficiency, so that NUE could be maximized and the application of nitrogen fertilizer could be reduced. It was found that less fertilization, no-tillage, and winter tillage were beneficial to the expression of AMF, such as the increase in sporulation rate, mycelial length, and root infection rate, and the combination of the three tillage measures had a greater impact on the expression of AMF. Different agricultural tillage measures have different effects on the growth and community distribution of AMF, and active tillage measures can be adopted in production practice to maintain the function of the agricultural ecosystem.

Among the fertilization measures, the effect of P fertilizer on AMF has been widely studied. The negative effects of high P on AMF have been confirmed by many studies and are widely recognized. The high content of P in soil inhibits the growth, reproduction, development, and functional expression of AMF, reduces the sporulation rate and mycelial growth of AMF, and decreases the community diversity of AMF. When the P concentration in the soil is low, the growth and development of AMF are favorable [32]. Applying P fertilizer in P-deficient soil improves the AMF community diversity in crop roots. In addition, a new generation of sequencing technology was used to compare the long-term effects of triple superphosphate and phosphate rock powder fertilization on the microbial community structure of the maize rhizosphere. The results showed that the richness of AMF was highest in the soil fertilized with phosphate rock powder [33], indicating that P has a significant effect on AMF. Furthermore, the content of N in soil also affects AMF, but the results regarding the effect of N application are inconsistent. Some studies have shown that a high nitrogen application rate has a negative effect on AMF, and increasing the amount of inorganic nitrogen application reduces AMF biomass, species richness, and diversity [34]. Other studies have found no significant effect [35]. Some studies have also found that N application increases the abundance of microbial functional groups and changes the composition of microbial communities [36]. When N is sufficient, soil organic carbon (SOC) accumulates, and the relative richness of fungal communities increases. The effect of fertilization on changes in AMF community composition may not be a direct effect but rather an indirect effect of changes in soil properties caused by fertilization. Additionally, Wang et al. showed that fertilization did not significantly affect the abundance and diversity of AMF [37] but altered the relative abundance and community composition of AMF in soil. Xu et al. also demonstrated this under low P fertilization conditions [38]. Regarding the comparison of the effects of organic fertilizer and inorganic fertilizer, some studies have shown that the number of AMF in paddy soil with long-term mixed application of rice straw and inorganic fertilizer is significantly higher than that in paddy soil treated with inorganic fertilizer alone or left unfertilized. Hu et al. studied the effect of a long-term fertilization system on the soil fungal community in the black soil area [39], indicating that the application of organic fertilizer is more beneficial than inorganic fertilizer for the fungal community. At the same time, it was pointed out that even if the fertilization system is the same, the fungal community structure of soil treated with identical fertilization may differ due to geographical variations. This reveals that the fertilization system should be considered in conjunction with geographical location and soil type.

The dependence of modern agriculture on pesticides makes their use and residue not only destroy the soil microenvironment but also pose a threat to the soil microbial community. High doses of pesticides do not affect the abundance of AMF but change the community composition of AMF. The types of pesticides commonly used in agriculture are fungicides, insecticides, and herbicides. Studies have shown that the response of AMF to pesticides varies with different types of pesticides. Fungicides are used to control pathogenic microorganisms, which have a significant negative impact on AMF, reducing spore germination and mycelial formation [40]; insecticides are used to control pests, which have no significant inhibitory effect on AMF [41] and can even stimulate AMF root infection and P absorption [42], overcoming the negative effects of pesticides on maize crop growth [43]; herbicides are used to control weeds, which can affect AMF both directly and indirectly by impacting the host plant of AMF. These studies suggest that pesticides do have an effect on AMF diversity, whether positive or negative.

2.2.3.2 Environmental factors

Understanding the effects of environmental changes on AMF and the response of plants is essential for predicting ecosystem responses at regional and global scales. Water status has a significant effect on the distribution of the AMF community. Precipitation, as a key factor affecting soil water content, has a great impact on microbial communities. The analysis of soil microbial community composition using the phospholipid fatty acid method showed that there were significant differences in AMF diversity in soils with different water contents. Compared with arid soils, the diversity of AMF was higher in soils with more precipitation [37]. The spatial and temporal changes in available water content have a direct or indirect effect on the community composition and distribution of AMF. Water changes can directly affect the infection and distribution of extraradical hyphae and can also indirectly affect AMF by influencing the composition of aboveground plant communities and soil properties. Sun et al. selected a grassland in Inner Mongolia for experimental research [44]. In July and August, the experimental site was subjected to regular and quantitative precipitation treatment. It was found that water conditions affected the formation of AMF spores in roots. With the increase in precipitation, the number of spores produced also increased, and the number of spores of Ambispora gerdemannii was higher, making it an indicative species of precipitation changes. In addition, the distribution and composition of the AMF community were significantly different under varying precipitation levels, indicating that soil water content is a driving factor for changes in the AMF community. A study on the effect of precipitation treatment on AMF community composition in the grasslands of northern China over seven consecutive years showed that the increase in precipitation had a positive effect on the biomass, spore density, mycelial density, and root infection rate of AMF, and it had a significant effect on the composition of AMF. This effect may be an indirect result of precipitation influencing soil pH and plant communities. Heinemeyer et al. showed that higher temperatures could promote the infection of AMF in host plants and the development of extra-root hyphae [45]. Hu et al. demonstrated that the hyphal length density (HLD) of AMF was positively correlated with the annual average temperature [46], and the annual average temperature was the main positive driving factor. Although the mycelial length density was positively correlated with the annual average temperature, there was no significant correlation between AMF root infection, the relative abundance of AMF, and the annual average temperature, indicating that temperature does not affect all indicators of AMF. However, some studies have shown that warming has a negative or neutral effect on AMF infection. Rillig et al. found that although temperature increases the abundance of extra-root hyphae [47], it reduces the content of glomalin-related soil protein (GRSP). Shi et al. showed that warming reduces the species richness of AMF but does not change the phylogenetic composition of AMF. Compared with precipitation, the increase in temperature and nitrogen content in soil had no direct and significant effect on AMF. Only when precipitation is sufficient do temperature and nitrogen play an active role in fungal diversity. When the soil is in a drought state, the soil microbial community does not respond significantly to temperature and nitrogen. Therefore, water seems to be the main limiting factor in the system of multiple environmental soil factors. The explanation given by some scholars is that water is very important for the diffusion and supplementation of nutrients in soil. Water stress places the soil microbial community in unfavorable growth conditions and may lead to the accumulation of nitrate or ammonium ions [48]. It may even reach the level of inhibition of some extracellular enzymes and suppress the activity of most microorganisms, resulting in the lack of response of microbial communities to climate warming [49]. However, the effects of various environmental factors on microbial community structure are not independent but form an organic whole. Therefore, when considering the effects of environmental factors, the combined effects of multiple factors should be taken into account.

2.2.3.3 Soil factors

SOC, nitrogen, soil moisture, and soil pH were significantly correlated with the diversity of AMF. Soil pH is considered to be a key driving factor affecting soil microbial abundance and community distribution, but the positive and negative driving effects are not conclusive. Dumbrell et al. found that soil pH was positively correlated with AMF diversity [50], which could directly affect the infection, reproduction, and community structure of AMF, and had a great impact on the microbial community. Xu et al. found that soil pH was negatively correlated with AMF abundance and phylogenetic diversity [51]. The diversity of AMF was also related to soil texture (sand, loam, and clay). Previous studies have shown that the soil particle size of sandy soil is larger and the aeration is good, which will promote the growth of AMF, indicating that the soil particle size distribution is negatively correlated with the growth of AMF. However, the results of Hu et al. are inconsistent with this [46], as it is considered that the increase in HLD is positively correlated with the soil particle size distribution. Lekberg et al. studied the changes in AMF community composition in maize roots planted in different soil textures through field experiments [52] and found that Gigasporaceae is more distributed in sandy soil. The distribution of Glomeraceae in clay is greater, indicating that the AMF community structure in soils of different textures is distinct. The size of soil aggregates also had a significant effect on the composition of farmland microbial communities, resulting in different responses of AMF under varying aggregate sizes. Different soil types (acid soil, calcareous soil, etc.) are also the main factors that determine the composition and species richness of soil AMF communities, and different AMF communities can be used to characterize soil types. In addition, different land-use types also affect the AMF community structure, and the AMF community structure of the two land-use types (farmland and grassland) is significantly different. The AMF diversity index and OTU number of farmland are significantly lower than those of grassland. Bainard et al. studied the AMF community in farmland [53], grassland, and pasture, and found that the diversity of AMF was significantly different. Soil fertility was negatively correlated with AMF diversity. The lower the soil organic matter content, the higher the relative abundance of AMF. Nitrogen and phosphorus fertilization led to a decrease in mycorrhizal biomass, which may be explained by the fact that in low-nitrogen and low-phosphorus soils, plants can allocate more carbon to AMF, thereby improving the growth and development of AMF and making the symbiotic relationship closer. The soil N/P ratio also has a certain regulatory effect on mycorrhizal symbiosis. Based on the experimental study of the cultivated land at the agricultural station in the black soil area of Northeast China, it was found that the change in soil physical and chemical properties had a significant effect on the diversity of AMF. The content of available phosphorus and organic matter in the soil was an important factor affecting the AMF community structure [50]. The AMF community structure was also affected by the content of potassium, calcium, and magnesium in the soil. Hu et al. found that among the soil factors, soil pH and soil particle size distribution were positively correlated with the length density of AMF hyphae [46], while SOC and soil-available nitrogen were negatively correlated with the length density of AMF hyphae. The root colonization rate and relative abundance of AMF also decreased with the increase in SOC content; that is, soil fertility was the main negative driving factor for AMF development.

2.2.3.4 Biological factors

: AMF is a type of obligate living biological nutrition, symbiotic microorganisms with plants, and is closely related to plants. Therefore, host plants and plant rhizosphere organisms affect the distribution of AMF species diversity. Van der Heijden et al. proposed as early as 1998 that AMF species diversity determines the species diversity of host plants and surrounding plant communities [54], which lays a foundation for the study of the influence of host plants on AMF diversity in the future. In grasslands and pastures with different host plants, the community composition of AMF is significantly different and exhibits a certain host preference, indicating that host plants are a related factor in the composition of AMF communities [55]. Additionally, the sporulation rate of fungi varies with host species, which further demonstrates the existence of host effects. Even within the same host species, different varieties also impact the composition of AMF and the root infection rate, as AMF infection varies among different varieties [56]. Mycorrhizal infection will also change with different combinations of plant–fungus. The effects of plants on the distribution of fungal communities can be influenced by mycorrhizal symbiosis, plant litter, and root exudates, which further affect the structural composition of AMF. In addition, vegetation coverage also has an effect on AMF, which is positively correlated with AMF diversity. It was found that the diversity of AMF was lower in areas with low vegetation coverage, which may be due to the decrease in carbon allocated to the underground when the total carbohydrate synthesized by plant photosynthesis decreased. The fungal community was affected by the change in net primary yield, resulting in the decrease of diversity. It is also partly due to the fact that soil moisture content affects aboveground productivity and further impacts the diversity of AMF [57]. AMF richness may also be influenced by changes in plant community composition. This shows that the influence of host plants on AMF is closely related to environmental factors and is the result of the interaction of various factors. In general, it is necessary to consider the habitat conditions of the community to study the effect of host plants on the AMF community structure. Rhizosphere organisms around the roots of host plants also have an effect on AMF. Plant growth-promoting rhizobacteria (PGPR) are a group of soil microorganisms that have a positive effect on plant growth and development, including many microbial groups, such as Pseudomonas fluorescens, Bacillus subtilis, Agrobacterium rhizogenes, and so on. They can infect the rhizosphere, root surface, and cell interior of plants. Both PGPR and AMF are beneficial to the sustainable development of agroecosystems. Most studies have pointed out the positive role of AMF and PGPR in promoting each other. PGPR promotes the germination and mycelial infection of AMF spores [58]. Soil animals, such as earthworms, springtails, and nematodes, also have an impact on the growth and development of AMF. On the one hand, the survival of AMF must depend on living plants because its ability to spread spores is not strong. The presence of soil fauna can potentially help AMF reach beyond the reach of the mycelium, which has a positive effect on AMF. On the other hand, the activities of soil animals may also destroy the external hyphae of AMF and its contact with the host roots, reducing the length and density of AMF hyphae, and thus having a negative impact on AMF. Gormsen et al. conducted an experiment on the effects of earthworms and collembolans on mycorrhizal fungi [59]. The results showed that earthworms and collembolans had little effect on the growth of AMF hyphae, inhibited spore formation, and reduced the infection rate and richness of AMF. The study by Ferreira et al. found a strong correlation between the abundance of plant-parasitic nematodes and AMF spores [60]. There was a negative correlation between them on the surface of rhizosphere soil, and the abundance of root-parasitic nematodes and soil AMF spores was also negatively correlated. This phenomenon may be caused by competitive exclusion between the two. Therefore, not only do host plants affect the abundance and community composition of AMF, but the rhizosphere organisms around them also have a significant influence.

3. Molecular basis of symbiosis between mycorrhizal fungi and the plant–soil system

With the development of society and the progress of science and technology, research in the molecular field of AMF has made continuous progress in both breadth and depth. Since the 1990s, research on AMF inoculation experiments and effect evaluation has gradually increased following the establishment of bacterial resource libraries in various locations. This research has mainly focused on describing new species, improving the classification system, and enhancing cultivation technology. With the continuous advancement of molecular sequencing technology, the molecular mechanism of AMF symbiosis with plants has become a research hot spot in the field of mycorrhizal studies. However, the mechanism of the symbiotic relationship between AMF and plants remains unclear. Based on existing research, it has been found that the symbiosis between AMF and plants can be divided into five stages: (1) identification before symbiosis; (2) the formation of contact and intrusion points; (3) invasion of the outer cortex; (4) the formation of arbuscular structures; and (5) the formation of vesicles and spores. After AMF infects plants, the plants induce AMF to release mycorrhiza (Myc) factors by secreting strigolactone signaling molecules from their roots. Myc factors induce the expression of host plant root-related genes through Ca^{2+} mediation in AMF cells [61], promote the mass reproduction of hyphae, and increase the germination of spores [62]. The formation of a large number of hyphae expands the contact area between plants and AMF, which facilitates signal exchange between AMF and plants, thereby inducing symbiosis between host plants and AMF. In the study of rice (*Oryza sativa* L.)–AMF symbiosis, Shi et al. found that the rice root phosphorus uptake pathway (direct uptake pathway) and the symbiotic phosphorus uptake pathway (symbiotic uptake pathway) were both regulated by the host plant phosphorus signaling network. At the same time, this study demonstrated that the transcription factor phosphate starvation response is a key regulator of AMF symbiosis with plants. The exchange of carbon sources and phosphorus nutrition is the core of plant–AMF symbiosis. In the plant–AMF symbiosis system, plants transmit about 20% of photosynthetic products to AMF for their growth. In return, AMF hyphae absorb nutrients such as phosphorus and nitrogen from the soil for host plants, promoting their growth and development [63]. Previous studies have suggested that sugar is the main form of carbon source nutrition provided by plants for AMF [49]. However, Jiang et al. [64] published a paper in science, which, for the first time, revealed that fatty acids are the primary form of carbon source transmitted by plants to mycorrhizal fungi during the symbiosis of AMF and plants, overturning the sugar theory presented in textbooks.

In plant–AMF symbionts, AMF endophytic hyphae invade the root cortex of plants and then develop into arbuscular structures, becoming the main site for material and information exchange between plants and AMF. After AMF infects plants, it can also form vesicle storage organs for storing lipids in root cells and produce a large number of spores in the soil, which can survive for many years. AMF external hyphae continuously develop in the soil, forming a large network of underground common mycorrhizal hyphae, helping the host absorb water and mineral elements from the soil, thereby improving the nutritional status of the host and enhancing the stress resistance of the plant. The hyphal network is a "highway" for the transport of substances (mineral elements, water, etc.) between plants and between AMF and plants. It can link plant roots of the same species or different species, different genera, and even different families. In recent years, it has been found that the AP2/ERF superfamily transcription factor WRI5a simultaneously activates its expression by targeting the AW-box element in the promoter of the plant fatty acid transporter STR and the phosphate transporter PT4 coding gene, thus acting as a molecular switch to initiate AMF symbiotic bidirectional nutrient exchange [65]. However, the synthesis, metabolism, and output of plant fatty acids are energy-consuming processes. During the continuous development stage of arbuscules, from branching and maturation to degradation, plants must dynamically adjust the rate of nutrient exchange in time and space according to their own nutritional status and needs, so as to maintain the stability of the AMF–plant symbiosis. The diversity of AMF and its symbiotic relationship with plants are driven by its genetic lineage and environmental factors [66]. The establishment and maintenance of the symbiotic relationship between AMF and plants largely depend on the supply of soil moisture and nutrients in the ecosystem [55]. In view of this, future studies on the symbiotic mechanism of AMF and plants should consider the effects of different soil nutrient gradients or variations to further reveal the AMF-mediated nutrient signal regulation process.

However, some studies have shown that inoculation with AMF has no effect on plant productivity or even inhibits its growth and development, reducing plant biomass [67], which makes people question the general promotion of host plant growth by AMF. Mortimer et al. believed that the increase in the respiration rate of AMF symbionts consumed part of the photosynthetic products of the host and inhibited the accumulation of biomass [68]. Xavier and Germida found that the increased assimilation of AMF in symbiosis with the host offset its respiratory consumption, and plant biomass did not change significantly [69]. Inspired by these comparative results, we believe that the effect of AMF on plant productivity may be related to AMF diversity, experimental control factors, geographical location, soil moisture, fertility, and quality.

The interaction between AMF and plants depends on the degree of response of plant growth and nutritional characteristics to the environment; that is, there is a trade-off relationship between plant and AMF symbiosis. When the host plant's income is greater than the input cost, it will promote the symbiosis between AMF and plants. When the cost is greater than the income, the host plant will weaken the symbiotic relationship with AMF. In a recent study, Zhang et al. revealed the molecular mechanism of the trade-off between AMF and plants. They found that the AP2/ERF protein ERM1/WRI5a-ERF12-TOPLESS transcription complex forms a self-regulated positive-negative feedback loop, which dynamically regulates nutrient exchange and arbuscular development processes, thereby maintaining the stability of mutualistic symbiosis. AP2/ERF proteins ERM1 and WRI5 a play a positive regulatory role in nutrient exchange in the early stage of mycorrhizal symbiosis and are redistributed in the later stage of symbiosis to avoid the excessive output of their own carbon sources. ERM1 and WRI5 moderately activate the expression of ERF12 in the later stage of symbiosis, thus stepping on the brake of the lipid transport process and preventing the ineffective output of plant resources.

4. Effects of AMF on stress resistance of the plant–soil system and its mechanism

As a high-quality microbial fertilizer, AMF can not only promote plant growth and biomass accumulation but also improve the resistance of host plants to environmental stress. According to existing research, we summarized the main mechanisms of AMF symbiosis to improve plant stress resistance: AMF activates systemic resistance mediated by plant hormones such as abscisic acid, brassinosteroids, gibberellins, and auxins in plants [70], improves plant antioxidant enzyme activity, regulates osmotic substances [71], enhances host photosynthesis [72], promotes nutrient absorption and utilization, maintains cell ion balance, and stimulates host plant growth. AMF secretes GRSP to improve soil structure and promotes the host to produce a large number of rhizosphere secretions to regulate the soil environment and enhance plant root structure. AMF can also improve plant stress resistance by competing for pathogen ecological sites. In addition, AMF has a synergistic effect with probiotics such as Rhizobium, Actinomycetes, and Brettanomyces naardensis, which can jointly promote the growth and development of host plants and improve soil quality [73], although its mechanism requires further study.

4.1 Improve plant drought resistance

Drought is one of the most harmful abiotic factors affecting plant growth and biomass accumulation. Long-term drought in arid and semi-arid regions of the world threatens the sustainability of agricultural ecosystems and affects the ecological service functions of other ecosystems. It has become a general consensus that AMF can improve the drought resistance and water-use efficiency of plants. Duan et al. found that inoculation with Glomus mosseae significantly increased the biomass, grain yield, and tiller number of wheat under drought conditions (40% field capacity) [74]. Kavadia et al. found that under drought stress (30% field capacity), inoculation of AMF could improve the ability of chickpea (*Cicer arietinum* L.) roots to absorb water and also regulate the redistribution of soil moisture by AMF hyphae, thereby improving the stress resistance of host plants. Studies have also shown that inoculation of AMF did not significantly affect the yield and water-use efficiency of maize and grass pea intercropping under sufficient soil moisture conditions, while AMF increased the yield, water-use efficiency, and nitrogen and phosphorus-use efficiency of maize and grass pea under drought conditions (32% field capacity).

AMF can not only directly absorb water but also promote the absorption of mineral elements, indirectly improving the drought resistance of plants. Recent studies have shown that under drought stress conditions (35% field capacity), all three AMFs improved the drought tolerance of tomato, and the growth-promoting effect of Rhizophagus intraradices was lower than that of *Rhizophagus irregularis* and *Glomus mosseae* [75]. It can be seen that the drought resistance conferred by AMF to plants is related to AMF strains. It is necessary to explore the mechanism of different AMF strains on plant stress resistance from morphological, physiological, metabolic, and molecular aspects.

At present, a large number of studies have been reported on the enhancement of plant drought tolerance by AMF, and the research scope involves morphology, physiology, biochemistry, and molecular aspects. The symbiosis between AMF and plants mainly improves the drought resistance of plants through the following seven aspects: (1) AMF promotes the increase of net photosynthetic rate and stomatal conductance of plant leaves, increases the longitudinal transpiration tension, and thus improves the water absorption power of plant roots; (2) AMF increases plant soluble proteins, amino acids, and other macromolecules. These substances, acting as osmotic protective agents, can reduce plant osmotic potential, increase antioxidant enzyme activity, and reduce proline accumulation and other N-containing substances caused by oxidative damage, thereby giving crops excellent drought resistance; (3) AMF promotes plants to secrete hormones that regulate the host root architecture, increase root length, density, diameter, and the number of lateral roots, and improve the water absorption capacity of plants; (4) AMF symbiosis allows extra-rhizospheric hyphae to expand beyond the plant rhizosphere nutrient and water depletion zone, enabling more effective absorption of water and nutrients, thereby improving stress resistance; (5) AMF can secrete and synthesize GRSP, which is conducive to the formation of soil macroaggregates and the improvement of soil structure, thereby enhancing the absorption of water and mineral elements by plants; (6) AMF can form a large number of mycelium bridges by connecting adjacent plants to optimize the allocation of resources between plants and alleviate the intraspecific or interspecific competition between adjacent plants [67]; and (7) AMF can regulate the expression of host genes and improve its drought resistance. For example, Singh et al. showed that in the intercropping system of pigeonpea (Cajanus cajan (L.) Druce) and millet (*Setaria italica* (L.) Beauv. var. germanica (Mill.) Schrad.) under drought stress, pigeonpea with deep roots hydraulically lifted the bottom water to the dry surface soil, and the shallow-rooted millet on the surface used hydraulic water through the AMF mycelium network to improve its survival rate. A new study found that AMF mediated a molecular regulatory mechanism of mycorrhizal plants in response to drought stress. Specifically, the AMF Heterorhizal cystomycetes transcriptional activator RiMsn2 regulates arbuscular mycorrhizal formation and controls the expression of drought stress response genes in mycorrhizal plants through the RiHog1-RiMsn2-STREs module, thereby enhancing the drought tolerance of host plants [76]. In summary, the defense responses of plants to drought stress in the symbiotic system are different, depending on the specific host plants and AMF species. Further work should be devoted to studying the effects of AMF dominant species in different regions on plant drought resistance. Secondly, research on AMF promoting plant stress resistance at home and abroad mainly focuses on the physiological and biochemical levels, and it still needs to be further explored at the molecular level.

4.2 Improve the ability of plants to resist saline-alkali

Soil salinization is a global problem affecting agricultural production and the ecological environment, and it is also one of the most important abiotic stresses affecting plant productivity. The area of saline-alkali cultivated land in China is as high as 9.21 million hm^2, accounting for 7% of the total cultivated land area in the country, and this value is increasing year by year. Under salt stress, plant cells are in a state of high osmotic potential, and their ability to absorb water and nutrients is reduced. High concentrations of salt ion stress not only limit plant growth and development but also cause specific toxicity to plants. AMF symbiosis can enhance the salt tolerance of host plants by promoting the growth of host plants, protecting the photosystem of host plants, enhancing the antioxidant system, regulating osmotic regulators, changing root architecture, and maintaining cell ion balance. Diao et al. found that moderate salt stress (100 mmol/L NaCl) inhibited the biomass accumulation of Suaeda salsa [77], and inoculation with AMF Glomus mosseae significantly increased the carbon and nitrogen accumulation of the host, thereby promoting biomass accumulation and improving its salt resistance. Liu et al. found in a recent study that inoculation of four mixed AMFs under moderate (100 mmol/L NaCl) and high (200 mmol/L NaCl) salt stress conditions can effectively increase the chlorophyll content and photochemical quenching coefficient of peanut (Arachis hypogaea L.), protect the host photosystem, and enhance the salt tolerance of peanut [78]. Studies have also shown that AMF symbiosis under moderate salt stress (100 mmol/L NaCl) increased the activities of antioxidant enzymes such as superoxide dismutase (SOD), peroxidase (POD), and catalase (CAT) in alfalfa (Medicago sativa L.), which was beneficial for removing excessive reactive oxygen species in host cells and reducing the level of membrane lipid peroxidation, thereby reducing the damage caused by high-concentration salt ions to the host. Under high salt stress (300 mmol/L NaCl), inoculation with AMF can regulate the content of osmotic regulators such as malondialdehyde, proline, soluble sugar, and soluble protein in Elaeagnus angustifolia L., thereby regulating the permeability of plant cells and improving their stress resistance [79]. In addition, under salt stress conditions, inoculation with AMF can increase the amount of free polyamines in plants. Under the regulation of AMF, polyamines can change the root configuration, mycorrhizal infection rate, and the number of arbuscules and vesicles in citrus (Citrus reticulata Blanco), thereby improving the salt tolerance of host plants. Jia et al. found that AMF colonization under salt stress (2–15 d S/m NaCl) significantly reduced the Na and Cl ion concentrations in Astragalus adsurgens Pall. plants and improved their salt tolerance [80]. Liu et al. showed that improved their salt tolerance [81].

There are few studies on the effect of AMF on the alkali resistance of host plants. Wang et al. inoculated maize with AMF Glomus mosseae under weak alkaline (pH 7.8) stress, which enhanced the stress resistance of maize by promoting its growth and development and increasing the levels of plant hormones and antioxidant enzymes [82]. Cartmill et al. [83] used mixed AMF (Glomus) to infect Catharanthus roseus L. at high bicarbonate (HCO_3^-) concentrations (7.5 g/mol and 10.0 g/mol). The results showed that AMF colonization significantly promoted the growth of *Catharanthus roseus* L. plants, increased antioxidant activity and photosynthetic rate, and improved the tolerance of the host to highly alkaline soil. AMF–plant symbionts are beneficial in alleviating the toxicity of saline-alkali to plants, improving the nutritional level of plants, maintaining cell ion balance, and enhancing the saline-alkali tolerance of host plants. Therefore, inoculation with AMF can alleviate the damage to plants under saline-alkali stress in a variety of ways, which has great potential value for the restoration of saline-alkali soil. In future research, it is necessary to focus on screening strains with rapid infection capabilities and strong resistance gene expression.

4.3 Increased resistance of plants to heavy metals

Soil heavy metal pollution has become a global concern, and farmland pollution is generally a combination of multiple heavy metals. Heavy metals with mobility can enter plants through soil, inhibiting plant growth and endangering human health [81]. AMF can reduce the absorption of heavy metals by plants and improve the absorption and utilization of mineral nutrients, so that plants show certain resistance to heavy-metal-contaminated soil, which is conducive to the growth of host plants and is of great significance to food security and the ecological environment [84]. By summarizing the results of previous studies, we found that the mechanism of AMF significantly improving plant productivity in heavy metal-contaminated soil mainly includes the following seven aspects: (1) improving photosynthesis to promote host growth and development; (2) promoting the absorption of nutrients; (3) activating the activity of antioxidant enzymes; (4) reducing the content of malondialdehyde and hydrogen peroxide; (5) inhibiting the transport of heavy metals from roots to shoots; (6) adsorbing heavy metals by mycelium; (7) upregulating a variety of metabolites (mainly flavonoids) to affect the interaction of metabolites–soil–microorganism, forming a core functional flora that resists heavy metals and promotes the nutrient cycle. Madejon et al. found that under heavy metal stress conditions, inoculation with AMF can enhance the absorption of nutrients and water by plants, enhance photosynthesis of leaves, promote plant growth, and increase plant biomass, thereby improving their ability to resist heavy metal stress [85]. Shen et al. showed that inoculation with AMF Funneliformis mosseae could promote the growth and development of maize in antimony (Sb)-contaminated soil, inhibit the accumulation of Sb in the upper part of maize, and play a biological dilution effect on heavy metal Sb, alleviating the toxicity of heavy metals to plants [86]. AMF-like genus Paraglomus sp. can adsorb and accumulate excessive lead (Pb) in the roots of Canna indica L. through its fungal structure (intercellular mycelia, intracellular mycelia, etc.), which reduces the toxic effect of Pb on plant cells and limits the migration of Pb. Wang et al. [87] inoculated AMF Glomus mosseae in the roots of alfalfa and found that AMF reduced the migration of cadmium (Cd), and $Cd2^{+}$ + was transformed into an organic-bound state. AMF increased the available nitrogen and phosphorus content, extracellular enzyme activity, organic matter content, and redox potential of rhizosphere soil, which could inhibit the transport of heavy metals from roots to aboveground parts. At the same time, AMF symbiosis upregulated 739 metabolites and led to changes in microbial communities to help plants resist heavy metal toxicity and promote nutrient cycling. In summary, many studies on AMF and plant resistance to heavy metal stress have been carried out at home and abroad, and gratifying achievements have been made. However, recent studies have shown that AMF inoculation of barley (Hordeum vulgare L.) plants planted in fly ash-improved soil can enrich soil heavy metals, which may significantly increase the amount of heavy metals entering the human food chain and cause great harm to agricultural soil and food security. This has called into question the "omnipotent" AMF, and the continuous exploration of the functional diversity and adaptability of AMF under different environmental conditions in future research will help to understand the practical application potential of AMF more comprehensively.

4.4 Improve the ability of plants to resist soil pathogens

Research reports on the relationship between AMF and various plant pathogens, such as fungal, bacterial, and nematode hazards, as well as the defense capabilities of AMF–plant symbiosis against many soil-borne microbial pathogens, are emerging. Studies have shown that inoculation of AMF can reduce the harm caused by Fusarium, Alternaria, Rhizoctonia, Bacillus, and other pathogens to host plants [88]. AMF can also enhance the host's resistance to foliar pathogens such as Magnaporthe oryzae. Symbiosis with AMF and the formation of mycorrhizae before the pathogen infects the plant is a prerequisite for AMF to induce plants to enhance their defense ability or reduce their susceptibility to the pathogen, and the effect of AMF depends on the diversity and richness of AMF, the infection potential and pathogenicity of soil pathogens, and other factors [89]. The different effects of AMF on pathogenic bacteria may be related to the similarity between pathogen infection and AMF colonization, that is, the trade-off between the symbiotic relationship between plants and AMF and plant resistance to pathogens. At present, AMF has been widely used in the biological control of various plant pathogens, such as crops, ornamental plants, and forest trees [90]. Understanding the mechanism of AMF disease resistance is of great significance for better application of AMF. Based on existing research, we believe that AMF colonization can improve plant disease resistance by promoting the growth of host plants (promoting nutrient absorption and improving photosynthesis), inducing the formation of plant disease resistance defense systems (regulating gene expression, activating plant hormones, and enhancing the antioxidant system), secreting root exudates to regulate the soil environment, improving plant root structure, and competing for pathogen ecological sites. For example, AMF addition can effectively promote the increase of rice biomass and grain yield, thereby enhancing its resistance to leaf pathogens such as rice blast. Et al. found that inoculation of AMF Glomus mosseae in the roots of poplar (Populus L.) seedlings effectively resisted the infection of *C. agglomerans*. AMF symbiosis increased the chlorophyll and soluble protein content of poplar leaves, significantly improved the level of photosynthesis, and enhanced disease resistance. The symbiosis between AMF and tea plants (*Camellia sinensis* (L.) O. Kuntze) mainly enhances their disease resistance by activating plant hormones such as abscisic acid, brassinosteroid, gibberellin, and auxin in tea plants. In addition, Wang et al. found that inoculation with AMF Glomus mosseae could significantly increase the content and activity of SOD, POD, and CAT in strawberry (Fragaria × ananassa Duch.), enhance the antioxidant capacity of the host, and thus reduce the incidence of Fusarium wilt. AMF can form a hyphal network in the host root system and change the morphological structure of the root system. At the same time, the hyphal network, as a natural physical and mechanical barrier, can effectively prevent the invasion of pathogenic bacteria into plant roots [90]. AMF and pathogens have the same sites on the root attachment sites of host plants, and they are competitive in spatial survival. The competitive attachment sites between AMF and pathogens can inhibit the binding of pathogens to host plants and reduce the infection rate of pathogens. When AMF successfully infects plant roots, it becomes difficult for pathogens to invade plants again, which inhibits the infection rate and reproduction

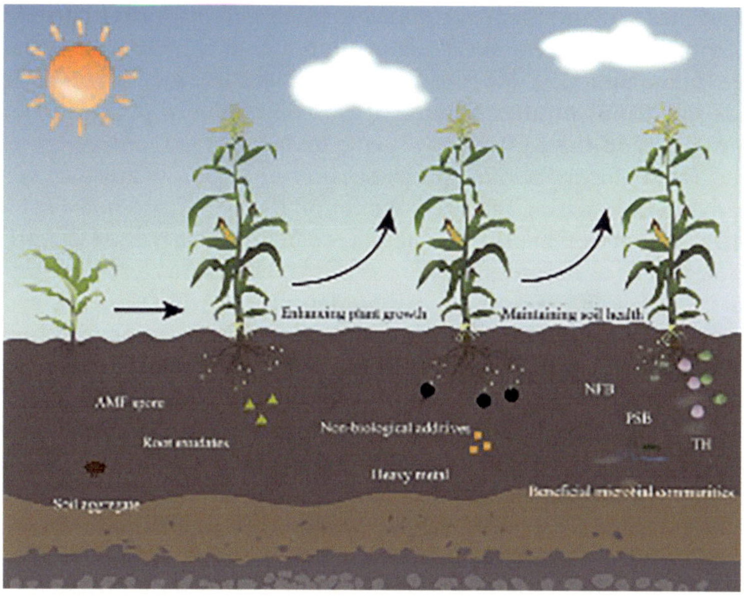

Figure 2.
The functional versatility of AMF.

rate of pathogens (**Figure 2**) [91]. On the contrary, when pathogens are dominant, AMF will be inhibited, thereby reducing disease resistance. In summary, research on AMF–plant symbiosis resistance to pests and diseases mainly focuses on the resistance response of plants induced by a single AMF genus (Glomus) or a single strain. In future research work and actual production activities, it is necessary to expand the combination of AMF and other beneficial microorganisms to control diseases and explore the combination of bacterial fertilizers suitable for different plant diseases.

5. Effects of mycorrhizal fungi on nutrient cycling in the plant–soil system and its mechanisms

AMF is a key driving factor that mediates the circulation of mineral elements between host plants and soil. Mycelia formed through symbiosis with plants can promote the efficient absorption and utilization of various nutrients in the soil by plants and enhance soil carbon sequestration by synthesizing GRSPs. This resource exchange forms the basis of the symbiotic relationship. AMF-driven global carbon, nitrogen, and phosphorus cycles are the most important material cycles in ecosystems. However, the mechanism of AMF uptake and utilization of mineral elements in plant–soil systems is not clear. Fig. 5 shows the main mechanism of AMF on nutrient cycling in the plant–soil system.

5.1 Mechanism of AMF symbiosis on soil organic matter

In ecosystems, soil carbon sequestration is often considered a potential way to mitigate global climate change. AMF is a key medium for plant photosynthetic carbon input into soil and mediates the absorption and utilization of soil mineral nutrients by plants. This symbiotic relationship, based on resource exchange, helps to improve soil quality [19]. Most of the previous studies focused on the organic compounds (glomalin) produced by AMF as a soil binder to promote the formation of soil aggregates and the fixation of organic matter, often ignoring the regulation of AMF-mediated biogeochemical processes on soil organic matter. In view of this, by summarizing the existing research, we found that AMF promotes soil carbon cycling and sequestration mainly through the following five ways: (1) AMF promotes the accumulation of underground biomass of plants and increases the input of soil organic matter through rhizosphere carbon deposition, which is a key source of soil organic matter; (2) AMF and plant symbiosis form a large number of hyphae to promote the host's absorption of mineral nutrients. At the same time, hyphae and plant roots form mycorrhizal structures to continuously adjust and bond soil particle aggregation, which helps the formation of soil macroaggregates and hinders the decomposition of soil organic matter; (3) AMF promotes soil carbon sequestration by competing with organic matter decomposers for soil nutrients and reducing organic matter decomposition. At the same time, AMF secretions can reduce soil pH, which is conducive to inhibiting the decomposition of soil organic matter; (4) The secretion of glomalin by AMF affects soil physical structure and drives the transport, redistribution, transformation, and storage of organic matter; and (5) AMF can significantly increase the activity of soil protease, urease, cellobiose hydrolase, etc., and significantly reduce the content of soil ammonium nitrogen, resulting in a decrease in microbial activity that decomposes organic matter, which is conducive to the storage of soil organic matter. At present, there are more and more studies on the effect of AMF on farmland soil quality under field conditions. Duan et al. found that in the arid and semi-arid areas of the Loess Plateau, field inoculation of AMF promoted the increase of soil organic matter and had great potential for the improvement of soil quality [92]. Tao and Liu studied 6,014 articles on the effects of AMF on SOC in a recent meta-analysis, and the results showed that AMF inoculation generally increased SOC stock and plant biomass accumulation [19]. Tian et al. showed that inoculation of AMF significantly increased the proportion of > 2 mm macroaggregates and soil enzyme activity in field soil compared with noninoculated AMF, indicating that inoculation of AMF is a promising soil carbon sequestration method [92]. AMF can significantly affect soil organic matter decomposition and nutrient cycling, which is conducive to improving soil quality and promoting sustainable agricultural development. In addition, mycorrhiza can improve the absorption and transport of nitrogen and phosphorus by host roots, and can also improve soil quality by enhancing soil aggregation and improving soil structure. However, AMF inoculation in fertile or acidic soils may deplete SOC [19]. The differences in the effects of AMF on soil carbon cycle and sequestration can be attributed to the diversity of AMF species, the characteristics of host plants, and the experimental conditions that affect the stability and decomposition of soil organic matter (such as soil nutrient level, pH value, and water content). Plant roots, initial soil SOC content, and pH are key factors affecting AMF-mediated SOC turnover [19]. Therefore, it is necessary to clarify the basic characteristics of soil, the abundance, diversity and infectivity of AMF propagules, so as to ensure the efficient ecological role of AMF in regulating soil mineral cycle and maintaining soil structure stability.

5.2 Mechanism of AMF symbiosis on soil phosphorus

Phosphorus is an indispensable element for plant growth and development, but most of the phosphorus in the soil exists in insoluble or organic forms, and cannot be directly absorbed and utilized by plants. AMF can form a symbiotic system with plant roots to provide a large amount of phosphorus and water for host plants, alleviate plant phosphorus deficiency, and promote plant growth and development. Phosphate-solubilizing bacteria are very important for the soil phosphorus cycle. These bacteria can decompose phosphorus in soil that is not used by plants into available phosphorus that can be utilized by plants. AMF improves soil phosphorus availability by hydrolyzing inorganic phosphorus and mineralizing organic phosphorus. It has been reported that AMF colonization significantly increases the number of phosphate-solubilizing bacteria and soil phosphatase activity, thereby promoting the increase of soil-available phosphorus content. Xie et al. found that the symbiosis of AMF and Chinese fir (*Cunninghamia lanceolata* L.) roots promoted the transformation of moderately decomposable organic phosphorus and primary mineral phosphorus into easily decomposable phosphorus, improving the effectiveness of soil phosphorus [93]. Inoculation of AMF in saline soil can activate insoluble phosphorus in soil into plant-available phosphorus and slowly available phosphorus, increasing soil phosphorus availability. Zhang et al. found that fructose and glucose in AMF mycelial secretions could stimulate the growth of phosphate-solubilizing bacteria and increase soil phosphatase activity, mainly because fructose stimulated the expression of phosphatase genes and secretion system-related genes of phosphate-solubilizing bacteria, thus improving soil phosphorus availability [94]. Jiang et al. found that the extracellular hyphae of AMF can stimulate and strengthen the enrichment and mineralization of organic phosphorus and the migration and transport of phosphate-solubilizing bacteria in organic phosphorus patches, thereby increasing the content of available phosphorus in soil [95]. The effect of AMF on plant growth is closely related to the improvement of plant phosphorus nutrition by mycorrhizal infection, especially in phosphorus-limited soils. Studies have shown that AMF hyphae can induce *Medicago truncatula* L. to secrete phosphatase to mineralize soil organic phosphorus, release gluconic acid to dissolve insoluble inorganic phosphorus, promote the absorption and transport of inorganic phosphate by hyphae in low-phosphorus soil, and further enhance the absorption of phosphorus by plants [96]. AMF colonization can also significantly improve the absorption of phosphorus in wheat seedlings, especially in low-phosphorus soil [97]. AMF colonization forms a large number of mycelial structures in the host root system, greatly increasing the absorption surface area of water and nutrient elements. A large number of hyphae can reduce the transport resistance of nutrients and effectively promote the transport of soil phosphorus to the host plant roots [12]. AMF contributes more to the formation of microbial communities, functional gene enrichment, and organic phosphorus mineralization than root hairs [98]. However, some studies have suggested that mycorrhiza has no effect on plant phosphorus uptake and growth, and sometimes even has an inhibitory effect. Reddell et al. [99] found in a pot experiment that when the soil phosphorus concentration was greater than 10 mg/kg, the AMF infection rate was significantly reduced, and when the phosphorus concentration exceeded 100 mg/kg, AMF could not successfully infect *Casuarina cunninghamiana* Miq., having no effect on host phosphorus absorption, growth, and development. Recent studies have found that with the increase of phosphorus concentration from 20 mg/kg to 100 mg/kg, the effect of AMF on the growth and phosphorus uptake of invasive plants and native plants changed from positive to negative [100]. Studies have also shown that when the soil phosphorus content is very low, AMF inoculation has an inhibitory effect on the phosphorus absorption of host plants [43, 101]. The above studies have led us to question the mycorrhizal promotion of mineral nutrient uptake and plant growth. In fact, the ability of AMF to promote plant absorption of soil phosphorus is limited to a very narrow concentration range (20–100 mg P/kg), and too high or too low soil-available phosphorus concentration will affect the absorption of phosphorus by AMF [43]. At the same time, studies have found that mycorrhizal plants are ubiquitous in phosphorus-deficient habitats, which may be due to the fact that mycorrhizal fungi improve the plant's defense against external environmental stresses, rather than simply promoting plant uptake of phosphorus. Therefore, the understanding of mycorrhiza on plant growth and development and mineral nutrient absorption and utilization also needs to move beyond the inherent theory and break the "cage."

The mechanism of AMF symbiosis with plants to promote the absorption of soil nutrients and water by plants mainly involves the following five points: (1) AMF and host plants form a large number of hyphae that interlace to create a network structure, which increases the contact area between host roots and soil, thereby improving the ability of plants to absorb soil moisture and nutrients [78]. The diameter of extracellular hyphae of AMF is usually 2–27 μm, which is smaller than the diameter of plant roots and can enhance the absorption and utilization of phosphorus by the host through extension [102]; (2) AMF promotes the secretion of acid phosphatase (such as oxalic acid, citric acid, and other acidic substances) by the roots of host plants. These substances can alter the pH of the rhizosphere soil of plants, thereby facilitating the transformation of mineral elements from the organic state, which is difficult for plants to utilize, to the effective state, which is easily absorbed [15]; (3) AMF symbiosis with plants can convert organic nitrogen in soil into arginine through the glutamine pathway, part of which is directly utilized by hyphae, while another part is decomposed into inorganic nitrogen and transported to plants to participate in the nitrogen cycle within plants [103]; (4) AMF can induce the expression of phosphorus transporter genes, enabling host plants to produce more phosphorus transporters [51]; and (5) AMF can increase the number of adventitious roots and lateral roots by promoting changes in the host root morphological configuration, ultimately enhancing the absorption of water and nutrients by plants [12, 15].

5.3 Mechanism of AMF symbiosis on soil nitrogen

Nutrient losses in agroecosystems threaten the global environment and human health, especially nitrogen losses. The low nitrogen content in farmland soil leads to nitrogen becoming a limiting element for plant growth and agricultural production. However, research on nitrogen utilization by AMF is relatively recent. The absorption and transport of nitrogen by AMF hyphae are caused by plant carbon exchange, which is similar to the reward mechanism of phosphorus and carbon exchange [27].

At present, mycorrhizal technology is widely used as a biological fertilizer in agricultural sustainable management and soil ecological restoration. The relationship between AMF and nitrogen needs to be clarified, and related research should receive more attention. Inorganic nitrogen is an important nitrogen source that plants can use, and nitrate nitrogen (NO^{-3}) and ammonium nitrogen (NH^{+4}) are two forms of inorganic nitrogen that plants can directly utilize. Cuenca and Azcón used *Erythrina variegata* L. as the experimental material; by supplying a single nitrogen source, NO-3, and inoculating AMF, the results showed that AMF symbiosis promoted the accumulation of biomass in Erythrina variegata. The nitrogen content of plants also increased, and the content of soil-mineralized nitrogen decreased. This study demonstrated that AMF promoted plants to utilize soil NO^{-3}. Studies by Duan et al. [104] and Johnson et al. [105] also demonstrated that AMF inoculation could enhance the uptake of soil inorganic nitrogen, especially NO^{-3}, by host plants. However, a recent study by Gou et al. showed that AMF inoculation in the soybean rhizosphere increased the abundance of host nitrogen fixation genes (36.9%) and nitrogen uptake [106]. Additionally, AMF addition also had the potential to promote soil particle solidification to regulate soil nitrogen loss.

In addition, AMF can cooperate with other beneficial microorganisms to promote nitrogen fixation and improve soil nitrogen limitation. The reasons for the inconsistent effects of AMF and plant symbiosis on soil nitrogen levels in the above studies may be due to the influence of different plant species and soil types. AMF colonization significantly increased the nitrogen content of maize (a nonnitrogen-fixing plant), while the content of soil inorganic nitrogen (NO^{-3}) and total nitrogen decreased significantly. After the symbiosis of nitrogen-fixing plants and AMF, AMF cooperates with nitrogen-fixing bacteria to promote biological nitrogen fixation and achieve the effect of fertilizing the field. For example, Yu et al. found that AMF and diazotrophs mainly formed a positive symbiotic relationship, which affected the biological nitrogen fixation process of the mangrove ecosystem by regulating the extracellular enzyme activity of mangrove rhizosphere microorganisms, changing the rhizosphere microenvironment of mangroves, improving the nitrogen fixation efficiency of the mangrove ecosystem, and alleviating soil nitrogen limitation [107]. A recent field study showed that maize symbiosis with AMF can absorb and transfer nitrogen fixed by its adjacent legume *Faidherbia albida* through mycelium, and the nitrogen transferred by AMF mycelium accounts for 33% of the nitrogen content in maize leaves. It can be seen that AMF has great potential to mediate plant uptake and transfer of leguminous plant-derived nitrogen and enhance the function of agricultural ecosystems.

Bender et al. used tomato mutants that could not be symbiotic with AMF and tomatoes that could be symbiotic with AMF as experimental materials under field cultivation conditions [108]. The study found that tomato plants symbiotic with AMF effectively inhibited soil NO^{-3} leaching, which may be due to AMF forming a mycelial network to increase the interception of soil nitrogen. At the same time, AMF also affects microbial activity involved in the nitrogen cycle. In addition, AMF colonization can not only utilize soil NO^{-3} but also absorb more NH^{+4}. Hawkins et al. found that the absorption of NH^{+4} by AMF Glomus mosseae was significantly higher than that of NO^{-3}, with the absorption of NH^{+4} being about 15 times that of NO^{-3}, and the absorption rate of NH^{+4} by AMF mycelium being significantly higher than that of NO^{-3}, which was due to the lower metabolic energy cost of absorbing and assimilating NH^{+4} [108]. AMF mycelium competes with nitrifying bacteria and denitrifying bacteria in soil for NO^{-3} and NH^{+4}, inhibits the action of nitrifying bacteria and denitrifying bacteria, and helps to reduce the emission of soil nitrous oxide (N_2O) [27], which is of great significance for improving the ecological environment and developing green agriculture.

Recent studies have shown that AMF Glomus mosseae mycelium can enrich denitrifying bacteria, especially Pseudomonas, and promote N_2O reduction, thereby reducing N_2O emissions from the stubble area of broad bean (*Vicia faba* L.). This study opens up new ways to develop sustainable agriculture and mitigate climate change by using cross-border microbial interactions.

More than 90% of the nitrogen in the topsoil (0–30 cm) exists in the form of organic nitrogen. AMF can absorb and utilize simple organic nitrogen such as glycine, glutamic acid, and arginine. The ability of AMF to absorb and utilize organic nitrogen varies, and the ability to absorb and transfer amino acids is lower than that of NH_4^+. Leigh et al. found that the formation of a large number of extraradical hyphae by AMF Glomus intraradices was beneficial to the utilization of soil organic nitrogen. In addition, studies on the symbiosis of AMF *Glomus clarum*, *Glomus claroideum,* and *Glomus* intraradices with Russian wild *Psathyrostachys juncea* (Fisch.) [109]. Nevski showed that these three AMF strains accelerated the mineralization of organic nitrogen in the hyphal chamber, which verified the conclusion that AMF could utilize soil organic nitrogen.

The AMF hyphal network drives soil organic nitrogen mineralization by stimulating microorganisms, and the mineralized nitrogen released in this process can be better utilized by plants. Especially in nitrogen-limited environments, the AMF hyphal network can provide additional nitrogen for host plants, thereby improving the competitiveness of the host for nutrient elements. AMF addition can reduce soil N_2O emissions by increasing plant nitrogen content and soil microbial biomass nitrogen, reducing soil-available nitrogen content, while inhibiting the microbial activity of decomposing organic matter, promoting the formation of soil macroaggregates, and increasing SOC sequestration, which has potential significance in alleviating global climate change.

The study of AMF inoculation on tomatoes under organic farm conditions showed that AMF promoted tomato growth and the absorption of nitrogen and phosphorus but had no effect on soil N_2O emissions. It can be seen that the effects of AMF on soil nitrogen utilization are often different under varying experimental conditions. At present, most of the research is carried out under controlled conditions, while the natural environment is complex and diverse. There are many kinds of AMF in soil, and the interaction between AMF strains may also affect the efficacy of AMF.

The AMF hyphal network drives soil organic nitrogen mineralization by stimulating microorganisms, and the mineralized nitrogen released in this process can be better utilized by plants. Especially in nitrogen-limited environments, the AMF hyphal network can provide additional nitrogen for host plants, thereby improving the competitiveness of the host for nutrient elements. AMF addition can reduce soil N_2O emissions by increasing plant nitrogen content and soil microbial biomass nitrogen, reducing soil-available nitrogen content, while inhibiting the microbial activity of decomposing organic matter, promoting the formation of soil macroaggregates, and increasing SOC sequestration, which has potential significance in alleviating global climate change.

The study of AMF inoculation on tomatoes under organic farm conditions showed that AMF promoted tomato growth and the absorption of nitrogen and phosphorus but had no effect on soil N_2O emissions. It can be seen that the effects of AMF on soil nitrogen utilization often vary under different experimental conditions. At present, most of the research is conducted under controlled conditions, while the natural environment is complex and diverse. There are many kinds of AMF in soil, and the interaction between AMF strains may also influence the effects of AMF.

6. Application of mycorrhizal fungi

6.1 Application of AMF technology in agriculture

Most crops show a certain mycorrhizal dependence on AMF. AMF can promote the absorption of soil nutrients by crops, enhance crop growth, increase crop yield, improve fertilizer-use efficiency, reduce fertilizer usage, and maintain sustainable agricultural development. At present, there are research institutions and companies around the world that can produce AMF inoculants on a large scale [110]. In practical applications, it is of great ecological and economic benefit to select crops with strong mycorrhizal dependence and high economic value for the application and promotion of mycorrhizal biotechnology. Many studies have shown that inoculation with AMF can increase the yield of major food crops such as maize, wheat (*Triticum aestivum*), and soybean (Glycine max), improve the nutritional status of crops, and enhance their resistance to drought, salinity, and other stresses [30, 31]. Using the commercial microbial agent MYKE PROSG2 (Canada) to inoculate maize, it was found that the phosphorus (P) content of host plants could still be increased in the presence of indigenous AMF [111]. Co-application of AMF and chemical fertilizers can improve the quality of rice and increase the content of iron (Fe) and zinc (Zn) in rice [33]. Ceballos' inoculation of cassava (Manihot esculenta), an important food crop in the tropics, with Rhizophagus irregularis has greatly increased its yield [28]. Each cassava plant is inoculated with 2,500 spores, which balances the effectiveness and economy of AMF inoculation.

6.2 Application of AMF technology in horticulture

Most horticultural crops can form a symbiotic system with AMF, such as Solanaceae, Allium, most fruit trees, herbaceous spices, Chinese herbal medicine, ornamental flowers, and so on. Horticultural crops have high economic value and are accustomed to using nursery and container seedlings, which provide convenient conditions for mycorrhizal plants at the seedling stage. After mycorrhizal inoculation, it can shorten the nursery period of seedlings, improve the survival rate of transplanting, promote the growth of seedlings, enhance disease resistance, and ultimately improve the yield and quality of products, increase the economic added value of horticultural crops, and obtain greater returns with less investment.

On the premise of reducing the amount of fertilizer by half, Meddad et al. [112] inoculated tomatoes with the commercial microbial agent Symbivit (France) and indigenous microbial agents *S. mosseae* and Septoglomus constrictum, which can increase plant biomass and achieve the effect of 100% fertilization without inoculation. By inoculating horticultural crops such as fruit trees and vegetables, the yield can be increased while the quality of fruits and vegetables can be improved. For example, the inoculation of Glomus trufemii on cucumber (Cucumis sativus) significantly increased the soluble protein and Vitamin C content in the fruit while increasing the single fruit weight and total yield per plant [35]. AMF can also promote the production of secondary metabolites and active substances in herbal spices and medicinal plants. For example, Salvia japonica was inoculated with commercial agents and Septoglomus viscosum. The content and quality of essential oils in the plants were significantly improved, and the proportion of laceol (an important component of a new biological pesticide) increased [36]. Xie et al. found that the contents of glycyrrhizic acid and liquiritin in the roots of Glycyrrhiza uralensis inoculated with AMF increased significantly [113]. Under adverse conditions such as drought and salinity, mycorrhizal plants have stronger adaptability, and the activities of antioxidant enzymes such as SOD, CAT, POD, and guaiacol peroxidase (GPOX) in the plant body are significantly increased. This can slow down the oxidative damage in plants and enhance their ability to cope with water deficits, resulting in stronger resistance [38, 39]. Selecting appropriate AMF to inoculate plants can not only promote the absorption of phosphorus (P) by plants but also increase the content of trace elements such as zinc (Zn), copper (Cu), and manganese (Mn) in plants [40, 41]. Therefore, mycorrhizal biotechnology can substantially reduce the use of chemical fertilizers and pesticides, avoid pesticide residues in fruit and vegetable products, and improve the quality of fruits and vegetables. In the production of fruit trees and flowers worldwide, AMF has been widely used as a biological protective agent, biological accelerator, and biological fertilizer. AMF plays an important role in the growth and nutrient absorption of fruit trees, as well as in improving fruit quality, stress resistance, and disease resistance. Because most of the root hairs of fruit trees are short and scarce, they are heavily dependent on the formation of reciprocal symbiosis with AMF to absorb nutrient elements. Therefore, it is of great significance to study the symbiotic relationship between fruit trees and AMF.

Studies on fruit trees have found that AMF can promote the growth of fruit trees by transporting mineral nutrients beyond the root range of fruit trees to the fruit trees [66]. In recent years, AMF has been widely used in apple (Malus domestica), pear (Pyrus pashia), bayberry (Myrica rubra), grape (Vitis vinifera), citrus (Citrus reticulata), and other fruit trees [7]. Therefore, on the basis of introducing the general situation of AMF research, this paper reviewed the application research progress of AMF in fruit tree growth, nutrient absorption, fruit quality, adversity physiology, and disease resistance at home and abroad in recent years, and discussed the problems existing in the current research and the future development direction, aiming to provide a reference for related research and promote the popularization and application of mycorrhizal technology.

6.3 Application of AMF technology in environmental remediation

6.3.1 Reclamation of mining area land abandoned after mining activities in the mining area

Generally, the growth of plants is limited due to poor soil structure, low content of organic matter, and mineral nutrients. The physiological and ecological characteristics of AMF make mycorrhizal technology an effective way to overcome the obstacles in the ecological reconstruction of mining areas. Abandoned or degraded land generally has low vegetation coverage, and AMF diversity and abundance are also low. The introduction of AMF inoculums or the enrichment of AMF by planting mycorrhizal-dependent species can improve the low AMF content in the local soil, thus exerting the ecological function of AMF.

In Inner Mongolia grassland, AMF inoculation of L. chinensis, a local dominant species, in sterilized soil can significantly increase the biomass and nutrient content of L. chinensis and improve the quality of forage [42]. AMF can also improve the soil rhizosphere enzyme activity, especially phosphatase activity [43], and enhance the soil rhizosphere micro-ecological environment. Bi et al. estimated that the use of mycorrhizal technology for the reclamation of mining areas could save 25,000 yuan per hectare of land, greatly reducing the cost of reclamation [114].

6.3.2 Heavy-metal-contaminated soil remediation

Many studies have shown that AMF can enhance the resistance of plants to various heavy metal pollution stresses and promote the survival and growth of plants in heavy metal-polluted environments. On the one hand, heavy metal toxicity causes the growth of plant roots to be hindered, making it difficult for plants to obtain mineral nutrients and water through root absorption. AMF can promote the absorption of mineral nutrients by plants through external hyphae, thereby improving plant mineral nutrition, promoting plant growth, and increasing plant biomass, which also dilutes heavy metals in plants and reduces heavy metal toxicity in plants [45–47]. On the other hand, AMF mycelia can effectively adsorb and immobilize heavy metals, accumulating heavy metals in fungi by binding sites of heavy metals in mycelia [48]. Chen et al. found that the mycelium of AMF can adsorb Mn, Zn, and Cd equivalent to 1.6%, 2.8%, and 13.3% of its own dry matter mass [115], and it can survive under heavy metal stress conditions through the avoidance mechanism. The effective immobilization of heavy metals by AMF can affect the distribution of heavy metals in plants and generally reduce the transport of heavy metals from roots to shoots [33, 45, 50]. This is also an important mechanism for AMF to protect host plants from heavy metal toxicity. For some heavy metals that can undergo valence and morphological transformation, such as arsenic (As) and chromium (Cr), AMF can also reduce the toxicity of heavy metals in plants by affecting the morphological transformation of heavy metals. X-ray absorption fine structure analysis based on synchrotron radiation light sources found that AMF can reduce highly toxic Cr (VI) to less toxic Cr (III) [51], and promote the absorption of S by plants, enhancing the detoxification ability of plants to Cr (VI) by upregulating the expression of high-affinity sulfate transporter genes [52]. Zhang et al. found that dimethylarsenic was present in the aerial parts of alfalfa (Medicago truncatula) and rice (Oryza sativa) grains infected by AMF [95, 96]. In addition, AMF can secrete organic acids, glomalin, and other substances, change the rhizosphere soil pH, redox potential, and other soil physical and chemical properties, and then affect the bioavailability of heavy metals in soil. For example, Spag-noletti et al. found that AMF is likely to inhibit the uptake of arsenic by plants by secreting more glomalin into the soil [54]. Glomalin fixes and isolates heavy metals such as lead (Pb), Cu, Zn, and As, reducing the toxicity of heavy metals to plants [56].

AMF can affect the absorption and accumulation of heavy metals by plants in many aspects and can effectively enhance the adaptability of plants to polluted environments. Therefore, it has broad application prospects in the ecological restoration of heavy-metal-contaminated soil. For the degraded soil in the mining area that needs revegetation, the application of AMF can quickly establish vegetation with species diversity and structural stability; for lightly polluted farmland, the accumulation of heavy metals in the aboveground parts or grains of crops can be controlled through the AM symbiosis system to maintain agricultural production and ensure food safety. In addition, heavy metal hyperaccumulators can also be combined for the bioremediation of contaminated farmland.

6.4 Influencing factors and applicable conditions of AMF application

In agricultural production, inoculation of exogenous AMF into crops can not only improve the absorption of nutrients by crops, especially nitrogen (N) and phosphorus (P), but also enhance soil structure, increase plant stress resistance, and inhibit the occurrence of crop diseases and insect pests. The study found that agricultural management methods affect the AMF community structure in the soil [57]. Planting intensity, tillage methods, and fertilizer application influence the abundance of AMF. The diversity of AMF in agricultural soil is low, and the fungal species are limited, usually dominated by several dominant species. Exogenous inoculation can alter the community structure of soil AMF to a certain extent. Combined with appropriate field management measures, such as adjusting chemical fertilizer application systems, implementing reasonable irrigation, reducing fallow periods, planting AMF-dependent crops, and practicing crop rotation, the abundance and diversity of soil AMF can be improved. In general, soil types with rare AMF and single community structures are ideal target soils, such as those subjected to frequent tillage, barren soils, reclaimed land in mining areas, and soil treated with various disinfection methods [1]. The effect of AMF inoculation is usually significant.

6.4.1 Host types and AMF species

Whether exogenous AMF can successfully infect the roots of host plants and establish a symbiotic relationship is determined by many factors, among which host plants and AMF species are the most important. It is more likely to be successful in AMF inoculation by selecting mycorrhizal-dependent host plants (such as legumes) [58]. On the contrary, planting nonmycorrhizal plants reduces AM propagules in the soil. When applying AMF in the field, we must pay attention to the planting of previous crops in cultivated land and adjust measures to local conditions.

AMF species that are not specific to the host plant and are easy to infect plant roots are often used when inoculating AMF, such as Rhizophagus intraradices, F. mosseae, and Rhizophagus heteromorpha [59]. Considering the differences in the adaptability of different AMF strains to soil, the current field soil inoculation generally uses a mixed microbial agent containing a variety of AMF for inoculation. Some microbial agent manufacturers will also choose a combination of multiple microorganisms to expand the scope of application of the microbial agent, such as AMF, rhizobium, and phosphobacteria combinations. However, the production cost and technical requirements of the compound microbial agent are high, and large-scale application is also limited.

6.4.2 Land-use patterns

Soil tillage usually destroys the mycelium network, which directly damages or kills AMF, while indirectly changing the soil environment, affecting the diversity and community distribution of AMF, and adversely affecting AMF. Thorough tillage significantly reduces the number of AMF spores and other propagules [60], so re-inoculation after tillage would reduce the abundance of indigenous AMF and increase the competitive advantage of exogenous AMF; although plowing after inoculation is beneficial to the spread of inoculum, it will reduce the success rate of inoculation [57]. Rotation usually enhances the number and diversity of AMF, and the more diversified the rotation form is, the more beneficial it is to improving the diversity and abundance of AMF [97].

6.4.3 Management measures

Irrigation methods affect the inoculation effect of AMF, and increasing the amount of water will reduce the infection rate of AMF while promoting the growth of plants. Moderate drought strengthens the growth-promoting effect of AMF on plants. The application of chemical fertilizers greatly increases the content of nutrients (especially N and P) in the soil in the short term, but it is unfavorable for AMF to infect plants. Therefore, when applying AMF agents, it is not appropriate to apply a large amount of chemical fertilizers, nor is it possible to simply mix the two. As an environmentally friendly bio-fertilizer, AMF can promote plants' absorption of soil nutrients and reduce the dependence of crop production on chemical fertilizers. It is generally not recommended to apply AMF inoculants to soils that are fertile or have high nutrient content due to large amounts of fertilization. In general, the use of organic fertilizer or slow-release fertilizer can promote the effect of AMF [98]. However, there are many kinds of organic fertilizers, which need to be selected and used according to the actual situation and cannot be applied excessively. At the same time, attention should be paid to issues such as pathogenic bacteria, heavy metals, and pH levels in organic fertilizers. In agricultural production, pesticides and fungicides are commonly used to control pests and diseases. Most of these chemicals are harmful to AMF. Herbicides also reduce the number and diversity of AMF in the soil while eliminating weeds. Therefore, chemical insecticides, fungicides, and herbicides should be avoided in the soil after the application of AMF, and physical methods or bio-friendly products should be used as much as possible to reduce toxicity to AMF. AMF agents must be used in conjunction with professional soil management models to fully realize their role.

7. Key scientific issues involved in the research and development of mycorrhizal application technology

7.1 Indigenous bacteria

AMF has strong ecological adaptability and has been found in various ecosystems on land, ranging from the Arctic to the Antarctic [62]. Indigenous AMF naturally exists in the soil for a long time, has adapted to the ecological conditions of the local soil, and possesses a survival advantage. Labidi et al. [99]

inoculated Hedysarum coronarium, a Mediterranean herbage, with commercial and indigenous microbial agents and found that indigenous AMF had a more significant effect on plant growth and increased phosphatase activity in the soil. Therefore, indigenous AM strains should be given priority in mycorrhizal applications. By planting mycorrhizal plants, combined with corresponding land management measures – such as reducing the application of chemical fertilizers and pesticides, and using organic or slow-release fertilizers – the enrichment of indigenous AMF can be achieved. While improving the effectiveness of indigenous AMF, this approach can also reduce costs, be easily accepted by farmers, and achieve sustainable use of soil. In addition, plant hormones such as strigolactones can be used as stimulants to encourage the infection of indigenous AMF on plants [26].

7.2 The viability, competitiveness, and long-term tracking and monitoring of exogenous AMF strains

The order in which microorganisms enter the soil, that is, the priority effect, affects the final microbial community composition in the soil [63]. Due to long-term artificial management of agricultural soil, the native AMF species are limited in diversity but high in abundance, and a stable community structure has been established. Therefore, exogenous AMF may have a strong competitive effect with indigenous AMF after entering the soil, resulting in the failure of inoculation [57]. Soil that has been abandoned for a long time, lacks plant growth, or is fertile has a low carrying capacity for AMF, making it difficult for exogenous AMF to be successfully inoculated. In field applications, it is necessary to screen efficient AMF for field production, and different AMF agents should be used in different habitats. Planting mycorrhiza and then transplanting it to the field soil environment is conducive to the formation and development of mycorrhiza [64, 65]. At the same time, increasing the inoculation amount of AMF and the probability of direct contact between AM propagules and roots can promote the infection of mycelium in plant roots and also facilitate the competition between exogenous AMF and indigenous AMF.

Pellegrino et al. carried out an in-situ study in the field and observed that the increase in crop yield was related to the intensity of AMF infection, and the mycorrhizal symbiosis system of exogenous inoculation could be maintained for two years [100]. Traditional staining methods are difficult to verify and quantitatively describe the symbiotic status of exogenously inoculated AMF and target crops. Although real-time quantitative PCR of nuclear ribosomal DNA (nrDNA) is becoming an effective tool for the quantitative determination of AMF biomass, it is still difficult to specifically amplify different AMF genotypes for this genomic region. Krak et al. designed a quantitative PCR system based on the large subunit of mitochondrial DNA (mtDNA) for the two ecotypes of Rhizophagus intraradices and accurately described the coexistence and competition of the two ecotypes in alfalfa roots. Current high-throughput sequencing technology provides a more powerful, sensitive, and quantitative method, which may be used to monitor the temporal and spatial dynamics of inoculated AMF and analyze whether exogenous AMF is affected by indigenous AMF communities. In addition, stable isotope probing can label and track the flow process of 13C in specific microbial communities, and it is possible to achieve long-term tracking and monitoring of AMF.

7.3 Stress degree and mycorrhizal dependence of plants

AMF must form mycorrhizal symbiotic structures with plants under certain environmental conditions to maximize its effectiveness. Many studies have shown that high nutrient levels inhibit AMF infection. Plants regulate the infection of AMF based on the supply level of nutrients in the environment. When the soil nutrient content is high, the infection rate is low, whereas when the soil nutrient content is low, the infection rate is relatively high. Soil-available phosphorus levels are the main influencing factor. Under high phosphorus (P) conditions, the P absorbed by plant roots can meet the needs of their own growth, and the mycorrhizal infection rate decreases. Excessive application of P fertilizer also reduces the distribution of photosynthetic products in plant roots, making the mycelium completely a consumer of photosynthetic products [68]. Plant roots may secrete a substance to inhibit AMF infection. However, when other nutrients in the environment, such as nitrogen (N) levels, are low, the adverse effects caused by high P may be weakened. Under medium N levels, AMF can play the greatest role, and the yield of tomatoes is the highest [101].

The establishment and development of mycorrhizal symbionts are affected by environmental factors, nutrient status, and other variables. Within a certain range, the more severe the environmental stress, the stronger the dependence of plants on mycorrhiza. Of course, there is not a simple linear correlation between the degree of stress and plant mycorrhizal dependence. Extreme environmental stress may lead to the inability of plants and AMF to adapt for survival. More systematic research is needed on the relationship between the degree of stress and mycorrhizal dependence of plants, as well as its universality.

7.4 Super bacteria

Some AM fungal species have a very wide geographical distribution and low environmental specificity. In this study, three genes from 82 isolates of AMF from six continents were sequenced and analyzed to study their geographical distribution, faunal characteristics, and the timing of migration with the continental plates. The results showed that the same genotype of AMF could be found on different continents, confirming that AMF species lack obvious geographical distribution. The formation and expansion of AMF species occurred later than the separation of Earth's plates, and human activities may have a significant impact on the dispersion and population structure of fungi. AMFs such as Rhizophagus intraradices, Rhizophagus irregularis, and Rhizophagus heterophyllus have no obvious specificity to host plants and are easily able to infect plant roots, resulting in a higher infection rate. At the same time, their growth-promoting effect on plants is usually more significant. Therefore, they are also known as "superbugs," which refers to strains with a wide application range and good application effect. However, in practical applications, the differences in the adaptability of different AMF strains to soil cannot be ignored, and applicable AMF strains can be flexibly selected according to plant types and habitat types.

AMF mixed inoculants are generally used in field inoculation to broaden the scope of application of inoculants. However, mixed inoculation relatively reduces the number of propagules of a single AMF strain and increases the spore density and infection rate of low-effect AMF, which may cause competition between different AMF strains. Therefore, the type and relative proportion of strain composition need to be considered when performing mixed inoculation of AMF. Some microbial inoculant manufacturers also choose the combination of AMF and beneficial microorganisms, such as rhizobia and phosphate-solubilizing bacteria, to improve the effectiveness of microbial inoculants, but the results of the application of mixed microbial inoculant products vary. Malusa et al. believed that the effect of mixed bacteria was better than that of single bacteria [102], but Herridge et al. believed that the effect of mixed bacteria was not better than that of single bacteria [103], and their production and commercialization face a series of technical problems.

7.5 Characterization of excellent strains–multiple stress resistance hypothesis

AMF can not only promote the absorption of mineral nutrients by plants but also enhance the resistance of plants to salinity, drought, diseases, and heavy metal pollution. AMF can improve the nutritional status of plants, regulate the balance of plant hormones, and improve the microenvironment of plant rhizosphere soil. This is an important basis for AMF to enhance plant stress resistance. Plants have their own phosphorus absorption channels, and AMF also has phosphorus transport genes, which can induce plants to produce specific phosphorus transport genes, thus promoting the absorption of P by plants. Under drought stress, the shoot phosphorus content of sorghum (Sorghum bicolor) inoculated with AMF was nearly doubled compared with that of noninoculated plants [73]. Zhang et al. [95] also believed that the absorption of P by mycorrhizal plants is an important mechanism to alleviate the toxicity of heavy metal Cd. In severely As-contaminated soil, the P concentration in mycorrhizal plants was two times higher than that in noninoculated plants, while the As concentration decreased, and the P/As ratio in mycorrhizal plants increased. The selective absorption of P and As by AM may play a key role in enhancing the As tolerance of plants. On the other hand, AMF usually contains higher phosphate groups, which may provide heavy metal binding sites. This plays an important role in immobilizing heavy metals in fungal structures, blocking the transfer of heavy metals from roots to shoots, and thus reducing the toxicity of heavy metals to plants. It can be seen that the AM symbiosis system with higher P absorption efficiency often has stronger heavy metal resistance and may also have stronger resistance to other stresses. The ability of AMF to absorb P can be used as a key indicator for screening excellent AMF strains. These excellent AMF strains often have multiple resistance capabilities.

The phosphorus uptake capacity of AMF may be affected by many factors, including host plant species, soil environment, and AMF species. From the perspective of AMF's own characteristics, the related biological traits that influence the activation, absorption, transport, and transmission of phosphorus will impact the absorption capacity of phosphorus: the ability of hyphae to secrete H^+ and phosphatase will affect the activation of soil phosphorus; the phosphate uptake capacity is determined by the density of extra-root hyphae and the type and quantity of phosphate transporters. The transport of phosphorus is greatly influenced by the infection rate, the connectivity of the hyphae in the root, and the synthesis of polyphosphate. The number of arbuscules and the number of phosphate transporters at the symbiotic interface determine the transport efficiency of phosphorus from AMF to plants. In general, the mycorrhizal symbiosis structure, especially the density of extra-root hyphae and the rate of intraroot arbuscular colonization, may be the key indicators affecting the efficiency of phosphorus nutrition in the AM symbiosis system and can be used as important indicators for screening excellent

AMF strains. A well-developed mycorrhizal symbiosis structure may indicate a higher mycorrhizal infection rate, but the high mycorrhizal infection rate and plant growth response are not simple linear correlations. This suggests that the excellent strains screened under certain conditions may not exhibit growth-promoting effects on plants under all conditions, as the effects of plant species and soil environmental conditions must also be considered. According to different application scopes and objectives, the special physiological characteristics of AMF should be comprehensively considered on the basis of basic physiological indexes. For example, the adaptability of AMF to polluted environments should be fully considered when screening AMF for application in contaminated soil remediation.

The benefits of the AM symbiosis system to plants are not limited to helping plants absorb mineral nutrients. For example, AMF can reduce the incidence and disease index of plants by promoting the formation of an anti-pathogen defense system in plants. In addition, AMF can also induce changes in plant morphology, physiology, biochemistry, gene expression, etc., improve the rhizosphere microenvironment of plants, and thus affect phytophagous insects. The antagonistic effects of AMF on pathogenic bacteria and herbivorous insects are different, and strict testing and evaluation are needed when screening biocontrol strains.

8. Conclusions and prospects

The development of mycorrhizal technology is inseparable from the support of basic research. As mentioned above, the related basic scientific issues involved in mycorrhizal application technology include the following: 1) Do excellent AMF strains have multiple stress resistance? What are the key indicators for identifying excellent strains? 2) The viability and competitiveness of exogenous strains and their key influencing factors? 3) The relationship between the degree of stress and plant mycorrhizal dependence? and 4) What is the relationship between AMF diversity and function? Which is better: "super bacteria" or mixed flora (indigenous flora)? These scientific problems are worthy of further systematic and in-depth study. Enterprises should expand the production efficiency of AMF, improve the quality of microbial agents, reduce the cost of microbial agents, and establish a sustainable market economy chain. Farmers should improve their awareness of ecological civilization construction and reduce the use of chemical fertilizers, as AMF fertilizer can help farmers obtain greater economic benefits. All parties should cooperate with each other and work together to promote the application of AMF technology on a wider scale and contribute to the construction of ecological civilization.

In order to comprehensively promote the development and wide application of mycorrhizal technology, the technical research and development directions that should be considered in the future include the following: 1) screening an index system of excellent strains suitable for different application ranges; 2) combining with actual production, developing different microbial agents to adapt to various application situations, such as the development of bio-organic fertilizer, soil conditioner, and seedling substrate based on mycorrhizal fungi; 3) developing mixed and mass production technology for different AMF strains and AMF combined with other beneficial microbial agents; 4) researching and developing specialized supporting inoculation technology for different crops, such as field crops, flowers, vegetables, and medicinal plants; 5) defining applicable conditions and scope of mycorrhizal technology, including applicable crop species, suitable soil and climate conditions, and applicable production systems, such as organic agriculture and high value-added farming; 6) developing indigenous mycorrhizal fungi effect stimulation techniques, such as signal substances and regulatory factors that promote mycorrhizal symbiosis; and 7) establishing supporting management technology for mycorrhizal technology, particularly water and fertilizer management, to form a comprehensive technology system. Without such a system, the advantages of mycorrhizal technology cannot be fully realized and may even be overshadowed.

There are still some bottlenecks in the application of mycorrhizal technology: 1) The pure culture technology of AMF needs to be further developed. Limited by the research progress in AMF basic biology, it is still impossible to obtain high-yield and high-quality microbial agents through fermentation technology. The traditional method of joint propagation with host plants can only temporarily meet the needs of scientific research, which is far from satisfying the requirements of large-scale field applications, and it is even more challenging to carry out genetic transformation at the molecular level. Relevant scientific research units should strengthen cooperation with enterprises to jointly explore ways to improve the quality and production efficiency of microbial agents, reduce product costs, and establish a sustainable market economy chain. 2) Many AMF agents and products performed well in environmentally controlled trials, but their effectiveness was significantly affected in the complex environment of field applications. The instability of this application effect is one of the most critical bottlenecks preventing AMF preparations from being applied on a large scale. A single technology is often insufficient to solve complex practical problems. The research and development of mycorrhizal technology should be more closely integrated with practical challenges, establishing a mycorrhizal application technology system that includes agent (compound product) production, suitability assessment, inoculation technology, and management-supporting technologies, rather than simply providing a single product or technology. If the practical application effect and benefits of microbial inoculant products cannot be effectively guaranteed, the research, development, application, and promotion of mycorrhizal technology will struggle to progress. 3) The public's awareness and acceptance of mycorrhizal technology remain very low, and it needs to be vigorously popularized and promoted. The inappropriate use of AMF reduces its application effectiveness and also greatly diminishes the enthusiasm of grassroots technology promotion personnel and farmers. The relatively high price of AMF further complicates its large-scale promotion. If the government and relevant departments implement certain price subsidies and provide technical support for mycorrhizal technology, they can help support and accelerate the rapid development of related industries. and 4) At present, there are no standardized quality criteria or application procedures for AMF inoculants. The current microbial fertilizer standards only stipulate the amount of bacteria, and these standards vary across different countries. Poor-quality inoculants reduce the effectiveness of mycorrhizal technology applications. Therefore, to play a greater role in agriculture, horticulture, and ecological restoration, AMF technology depends on the joint efforts of government departments, scientific research units, enterprises, and grass-roots staff. It is predicted that the global population will exceed 9 billion [76] by 2050, and the world faces severe resource and environmental crises. Mycorrhizal technology is an environmentally friendly biotechnology. The use of AMF as a biofertilizer in agriculture and horticulture can effectively improve the yield and quality of agricultural products, increase farmers' income, reduce the use of chemical fertilizers and pesticides, and minimize environmental pollution. In the field of ecological restoration, mycorrhizal technology can also be applied to the rehabilitation and treatment of degraded ecosystems and polluted environments. In the context of increasingly serious resource and environmental challenges, mycorrhizal technology must have broader development potential and application prospects.

Conflict of interest

The authors declare no conflicts of interest.

Author details

Ying Wu Shi[1,2,3*], Hongmei Yang[1,2,3], Ming Chu[1,2,3], XinXiang Niu[3,4], Huifang Bao[1,2], Ning Wang[1,2,3], Faqiang Zhan[1,2], Xuanqi Long[1,2], Rong Yang[1,2], Yun Chen[1,5], Shuang Dou[1,6], Zhichao Meng[1,6], Zhao Zhang[1,6] and Kai Lou[1,2]

1 Institute of Microbiology, Xinjiang Academy of Agricultural Sciences, Xinjiang, China

2 Xinjiang Laboratory of Special Environmental Microbiology, Xinjiang, China

3 Key Laboratory of Agriculture Environment in Northwest Oasis of Agriculture, Xinjiang, China

4 Institute of Soil and Fertilizer and Agricultural Water Conservation, Xinjiang Academy of Agricultural Sciences, Xinjiang, Urumqi, China

5 College of Agronomy, Xinjiang Agricultural University, Xinjiang, China

6 College of Life and Science and Technology, Xinjiang University, Xinjiang, China

*Address all correspondence to: syw1973@126.com

References

[1] Bunn RA, Corrêa A, Joshi J, Kaiser C, Lekberg Y, Prescott CE, Sala A, Karst J. What determines transfer of carbon from plants to mycorrhizal fungi? New Phytologist. 2024;**244**(4): 1199–1215

[2] Wang WH, Hu YJ, Wang XS, Zhao HH, Liu RJ. Investigations on arbuscular mycorrhizal fungi in the rhizosphere of wheat grown in Qingdao. Journal of Qingdao Agricultural University. 2017;**34**(2):90–94

[3] Gao CM, Liu N, Wang HB, Liu RJ, Li M. Arbuscular mycorrhizal fungal community structure in apple orchards of eastern Shandong. Journal of Qingdao Agricultural University. 2014;**31**(2): 79–84.

[4] Su YY, Guo LD. Arbuscular mycorrhizal fungi in non-grazed, restored and over-grazed grassland in the Inner Mongolia steppe. Mycorrhiza. 2007;**17**:689–693

[5] Yuan T, Tao GY, Jiang L. Arbuscular mycorrhizal fungi in the rhizospheric soil of four forest types in Fanjingshan National Nature Reserve. Journal of Northeast Forestry University. 2018;**46** (3):83–86

[6] Liu M, Zheng R, SL B, Qin Y, XW J. Seasonal dynamics of arbuscular mycorrhizal fungal community in the rhizosphere of Clematis fruticosa in the Daqingshan Mountains, Inner Mongolia. Acta Botanica Boreali-Occidentalia Sinica. 2016;**36**(9): 1891–1899

[7] Cai XB, Peng YL. Geographical distribution of arbuscular mycorrhizal fungi in the grass-land ecosystems of the Tibetan Plateau. Acta Ecologica Sinica. 2016;**36**(10):2807–2818

[8] Ahammed GJ, Hajiboland R. Introduction to Arbuscular Mycorrhizal Fungi and Higher Plant Symbiosis: Characteristic Features, Functions, and Applications. In Arbuscular Mycorrhizal Fungi and Higher Plants: Fundamentals and Applications. Singapore: Springer Nature Singapore; 2024. p. 1–17

[9] Selosse MA, Rousset F. The plant-fungal marketplace. Science. 2011;**333**:828–829

[10] Zhu X, Song F, Liu S, Liu F. Arbuscular mycorrhiza improve growth, nitrogen uptake, and nitrogen use efficiency in wheat grown under elevated CO2. Mycorrhiza. 2016;**26** (2):133–140

[11] Zhu X, Song F, Liu S, Liu F. Role of arbuscular mycorrhiza in alleviating salinity stress in wheat (*Triticum aestivum* L.) Grown Under Ambient and Elevated CO 2. Journal of Agronomy and Crop Science. 2016;**202** (6):486–496

[12] Malhi Y, Franklin J, Seddon N, Solan M, Turner MG, Field CB, Knowlton N. Climate change and ecosystems: Threats, opportunities and solutions. Philosophical Transactions of the Royal Society B. 2020;**375**(1794)

[13] Newbold T, Adams GL, Albaladejo Robles G, Boakes EH, Braga Ferreira G, Chapman ASA, Etard A, Gibb R, Millard J, Outhwaite CL, Williams JJ. Climate and land-use change homogenise terrestrial biodiversity, with consequences for ecosystem functioning and human well-being.

Emerging Topics in Life Sciences. 2019;**3**(2):207–219

[14] Smith SE, Read D. Mycorrhizal Symbiosis. London: Academic Press; 2008

[15] Tang D, Yan L, Fisher JL, Kang H, Dallongeville P, Wasaki J, Wang X, Lambers H. Why are some invasive plant species so successful in nutrient-impoverished habitats in south-western Australia: A perspective based on their phosphorus-acquisition strategies. Functional Ecology. 2025;**39**(2):635–652

[16] Hu XY, Wu HY, Zhu CS, Li W, Guo X, Li M, Guo SX. Effects of arbuscular mycorrhizal fungi and their dominant species on Flaveria bidentis plant growth and development. Mycosystema. 2023;**42**(3): 692–706

[17] Zhao HB, Zheng H, Miao CH, Shao TT, Feng YB.[Spatial-temporal pattern and factor diagnoses of agroecosystem health in major grain producing areas of Northeast China: A case study in Jilin Province]. Chinese Journal of Applied Ecology. 2016;**27**(10):3290–3298

[18] Redecker D, Schuessler A, Stockinger H, Stürmer SL, Morton JB, Walker C. An evidence-based consensus for the classification of arbuscular mycorrhizal fungi (Glomeromycota). Mycorrhiza. 2013;**23**:515–531

[19] Turrini A, Agnolucci M, Palla M, Tome E, Tagliavini M, Scandellari F, Giovannetti M. Species diversity and community composition of native arbuscular mycorrhizal fungi in apple roots are affected by site and orchard management. Applied Soil Ecology. 2017;**116**:42–54

[20] Mao JH, Ning DL, Li RB, Jing YB. Arbuscular mycorrhizas in three major olive growing regions of Yunnan Province. Journal of West China Forestry Science. 2015;**44**(1):114–120

[21] Oehl F, Laczko E, Oberholzer H-R, Jansa J, Egli S. Diversity and biogeography of arbuscular mycorrhizal fungi in agricultural soils. Biology and Fertility of Soils. 2017;**53**:777–797

[22] Zhou X, Tian L, Zhang J, Ma L, Li X, Tian C. Rhizospheric fungi and their link with the nitrogen-fixing Frankia harbored in host plant Hippophae rhamnoides L. Journal of Basic Microbiology. 2017;**57**:1055–1064

[23] Fierer N, Jackson RB. The diversity and biogeography of soil bacterial communities. Proceedings of the National Academy of Sciences of the United States of America. 2006;**103**:626–631

[24] Hu W, Strom N, Haarith D, Chen S, Bushley KE. Mycobiome of cysts of the soybean cyst nematode under long term crop rotation. Frontiers in Microbiology. 2018;**9**:386

[25] Schultz PA, Bever JD, Morton JB. Acaulospora colossica sp. nov. from an old field in North Carolina and morphological comparisons with similar species, A. laevis and A. koskei. Mycologia. 1999;**91**:676–683

[26] Zhang C, Liu G, Song Z, Qu D, Fang L, Deng L. Natural succession on abandoned cropland effectively decreases the soil erodibility and improves the fungal diversity. Ecological Applications. 2017;**27**: 2142–2154

[27] Chaudhary VB, Lau MK, Johnson NC. Macroecology of Microbes-

biogeography of the Glomeromycota, Ajit V ed. Springer: Mycorrhiza. Berlin; 2008. pp. 529–563

[28] Brito I, Goss MJ, De Carvalho M, Chatagnier O, Van Tuinen D. Impact of tillage system on arbuscular mycorrhiza fungal communities in the soil under Mediterranean conditions. Soil and Tillage Research. 2012;**121**: 63–67

[29] Jie W, Liu X, Cai B. Diversity of rhizosphere soil arbuscular mycorrhizal fungi in various soybean cultivars under different continuous cropping regimes. PLoS One. 2013;**8**(8)

[30] Hu J, Yang A, Zhu A, Wang J, Dai J, Wong MH, Lin X. Arbuscular mycorrhizal fungal diversity, root colonization, and soil alkaline phosphatase activity in response to maize-wheat rotation and no-tillage in North China. Journal of Microbiology. 2015;**53**:454–461

[31] Gao P, Yan FY, Meng C, Li F, Duan TY. Diversity of arbuscular mycorrhizal fungi under different agricultural practices in Loess Plateau in China. Pratacultural Science. 2016;**33** (10):1917–1923

[32] Verzeaux J, Hirel B, Dubois F, Lea PJ, Tetu T. Agricultural practices to improve nitrogen use efficiency through the use of arbuscular mycorrhizae: Basic and agronomic aspects. Plant Science. 2017;**264**:48–56

[33] Silva UC, Medeiros JD, Leite LR, Morais DK, Cuadros-Orellana S, Oliveira CA, de Paula Lana UG, Gomes EA, Dos Santos VL. Long-term rock phosphate fertilization impacts the microbial communities of maize rhizosphere. Frontiers in Microbiology. 2017;**8**:1266

[34] Albizua A, Williams A, Hedlund K, Pascual U. Crop rotations including ley and manure can promote ecosystem services in conventional farming systems. Applied Soil Ecology. 2015;**95**:54–61

[35] Tian H, Drijber RA, Zhang JL, Li XL. Impact of long-term nitrogen fertilization and rotation with soybean on the diversity and phosphorus metabolism of indigenous arbuscular mycorrhizal fungi within the roots of maize (*Zea mays* L.). Agriculture Ecosystems and Environment. 2013;**164**:53–61

[36] Zhang N, Wan S, Guo J, Han G, Gutknecht J, Schmid B, Yu L, Liu WX, Bi J, Wang Z, Ma KP. Precipitation modifies the effects of warming and nitrogen addition on soil microbial communities in northern Chinese grasslands. Soil Biology and Biochemistry. 2015;**89**:12–23

[37] Wang X, Sui X, Liu Y, Xiang L, Zhang T, Fu J, Li AR, Yang PZ. N-P fertilization did not reduce AMF abundance or diversity but alter AMF composition in an alpine grassland infested by a root hemi- parasitic plant. Plant Diversity. 2018;**40**: 117–126

[38] Xu J, Liu S, Song S, Guo H, Tang J, Yong JW, Ma Y, Chen X. Arbuscular mycorrhizal fungi influence decomposition and the associated soil microbial community under different soil phosphorus availability. Soil Biology Biochemistry. 2018;**120**:181–190

[39] Hu X, Liu J, Wei D, Hu X, Liu J, Wei D, Zhu P, Cui XA, Zhou B, Chen X, Jin J, Liu X, Wang G. Effects of over 30-year of different fertilization regimes on fungal community com- positions in the

black soils of northeast China. Agriculture, Ecosystems Environment. 2017;**248**:113–122

[40] Zocco D, Van Aarle IM, Oger E, Lanfranco L, Declerck S. Fenpropimorph and fenhexamid impact phosphorus translocation by arbuscular mycorrhizal fungi. Mycorrhiza. 2011;**21**:363–374

[41] Schweiger PF, Jakobsen I. Dose-response relationships between four pesticides and phosphorus uptake by hyphae of arbuscular mycorrhizas. Soil Biology and Biochemistry. 1998;**30**: 1415–1422

[42] Spokes JR, Macdonald RM, Hayman DS. Effects of plant-protection chemicals on vesicular-arbuscular mycorrhizas. Pesticide Science. 1981;**12**:346–350

[43] Corkidi L, Bohn J, Evans M. Effects of bifenthrin on mycorrhizal colonization and growth of corn. HortTechnology. 2009;**19**:809–812

[44] Sun X, Su Y, Zhang Y, Wu M, Zhang Z, Pei K, Sun L, Wan S, Liang Y. Diversity of arbuscular mycorrhizal fungal spore communities and its relations to plants under increased temperature and precipitation in a natural grassland. Chinese Science Bulletin. 2013;**58**:4109–4119

[45] Heinemeyer A, Fitter AH. Impact of temperature on the arbuscular mycorrhizal (AM) symbiosis: Growth responses of the host plant and its AM fungal partner. Journal of Experimental Botany. 2004;**55**:525–534

[46] Hu Y, Rillig MC, Xiang D, Hao Z, Chen B. Changes of AM fungal abundance along environmental gradients in the arid and semi-arid grasslands of northern China. PLoS One. 2013;**8**(2)

[47] Rillig MC, Wright SF, Shaw MR, Field CB. Artificial climate warming positively affects arbuscular mycorrhizae but decreases soil aggregate water stability in an annual grassland. Oikos. 2002;**97**:52–58

[48] Stursova M, Crenshaw CL, Sinsabaugh RL. Microbial Responses to Long-Term N Deposition in a Semiarid Grassland. Microbial Ecology. 2006;**51**:90–98

[49] Sheik CS, Beasley WH, Elshahed MS, Zhou X, Luo Y, Krumholz LR. Effect of warming and drought on grassland microbial communities. ISME Journal. 2011;**5**:1692–1700

[50] Dumbrell AJ, Nelson M, Helgason T. Dytham C,Fitter AH. Relative roles of niche and neutral processes in structuring a soil microbial community. The ISME Journal. 2010;**4**:337–345

[51] Xu T, Veresoglou SD, Chen Y, Rillig MC, Xiang D, Ondřej D, Hao Z, Liu L, Deng Y, Hu Y, Chen W, Wang J, Jizheng H, Chen B. Plant community, geographic distance and abiotic factors play different roles in predicting AMF biogeography at the regional scale in northern China. Environmental Microbiology Reports. 2016;**8**:1048–1057

[52] Lekberg Y, Koide RT, Rohr JR, Laura ALDRICH-WOLFE, Morton JB. Role of niche restrictions and dispersal in the composition of arbuscular mycorrhizal fungal communities. Journal of Ecology. 2007;**95**:95–105

[53] Bainard LD, Chagnon PL, Cade-Menun BJ, Lamb EG, LaForge K, Schellenberg M, Hamel C. Plant communities and soil properties mediate agricultural land use impacts on arbuscular mycorrhizal fungi in the Mixed Prairie ecoregion of the North American Great Plains. Agriculture, Ecosystems Environment. 2017;**249**:187–195

[54] van der Heijden MGA, Klironomos JN, Ursic M, Moutoglis P, Streitwolf-Engel R, Boller T, Wiemken A, Sanders IR. Mycorrhizal fungal diversity determines plant biodiversity, ecosystem variability and productivity. Nature. 1998;**396**:69–72

[55] Bao XZ, Wang YT, Olsson PA. Arbuscular mycorrhiza under water-carbon phosphorus exchange between rice and arbuscular mycorrhizal fungi under different flooding regimes. Soil Biology Biochemistry. 2019;**129**:169–177

[56] Brito I, Goss MJ, De Carvalho M, Chatagnier O, Van Tuinen D. Impact of tillage system on arbuscular mycorrhiza fungal communities in the soil under Mediterranean conditions. Soil and Tillage Research. 2012;**121**:63–67

[57] Barto EK, Rillig MC. Does herbivory really suppress mycorrhiza? A meta-analysis. Journal of Ecology. 2010;**98**:745–753

[58] Artursson V, Finlay RD, Jansson JK. Interactions between arbuscular mycorrhizal fungi and bacteria and their potential for stimulating plant growth. Environmental Microbiology. 2006;**8**:1–10

[59] Gormsen D, Olsson PA, Hedlund K. The influence of collembolans and earthworms on AM fungal mycelium. Applied Soil Ecology. 2004;**27**:211–220

[60] Ferreira BS, Santana MV, Macedo RS, Silva JO, Carneiro MA, Rocha MR. Co-occurrence patterns between plant-parasitic nematodes and arbuscular mycorrhizal fungi are driven by environmental factors. Agriculture, Ecosystems Environment. 2018;**265**:54–61

[61] He JM, Zhang C, Dai HL, Liu H, Zhang XW, Yang J, Chen X, Zhu YY, Wang DP, Qi XF, Li WC, ZH W, An GY, Yu N, ZH H, Wang YF, Xiao YL, Zhang P, Wang ET.A Lys M receptor heteromer mediates perception of arbuscular mycorrhizal symbiotic signal in rice. Molecular Plant. 2019;**12** (12):1561–1576

[62] Song FQ, Wang L, Ma F. Study on Arbuscular Mycorrhizal fungi-Amorpha Fruticosa Symbiosis System. Beijing: Science Press; 2013

[63] Chen BD, Yu M, Hao ZP, Xie W, Zhang X. Research progress in arbuscular mycorrhizal technology. Chinese Journal of Applied Ecology. 2019;**30**(3):1035–1046

[64] Jiang YN, Wang WX, Xie QJ, Liu N, Lx L, Wang DP, Zhang XW, Yang C, Chen XY, Tang DZ, Wang ET. Plants transfer lipids to sustain colonization by mutualistic mycorrhizal and parasitic fungi. Science. 2017;**356** (6343):1172–1175

[65] Jiang YN, Xie QJ, Wang WX, Yang J, Zhang XW, Yu N, Zhou Y, Wang ET. Medicago AP2-domain transcription factor WRI5a is a master regulator of lipid biosynthesis and transfer during mycorrhizal symbiosis. Molecular Plant. 2018;**11**(11): 1344–1359.

[66] Albornoz FE, Dixon KW, Lambers H. Revisiting mycorrhizal dogmas: Are mycorrhizas really functioning as they are widely believed to do? Soil Ecology Letters. 2021;**3**(1):73–82

[67] Nacoon S, Ekprasert J, Riddech N, Mongkolthanaruk W, Jogloy S, Vorasoot N, Cooper J, Boonlue S. Growth enhancement of sunchoke by arbuscular mycorrhizal fungi under drought condition. Rhizosphere. 2021;**17**

[68] Mortimer PE, Perez-Fernandez MA, Valentine AJ. The role of arbuscular mycorrhizal colonization in the carbon and nutrient economy of the tripartite symbiosis with nodulated Phaseolus vulgaris. Soil Biology and Biochemistry. 2008;**40**(5): 1019–1027

[69] Xavier LJC, Germida JJ. Selective interactions between arbuscular mycorrhizal fungi and Rhizobium leguminosarum bv. viciae enhance pea yield and nutrition. Biology and Fertility of Soils. 2003;**37**(5):261–267

[70] Liu CY, Hao Y, Wu XL, Dai FJ, Abd-Allah EF, Wu QS, Liu SR. Arbuscular mycorrhizal fungi improve drought tolerance of tea plants via modulating root architecture and hormones. Plant Growth Regulation. 2024;**102**(1):13–22

[71] Wang Y, Zou YN, Shu B, Wu QS. Deciphering molecular mechanisms regarding enhanced drought tolerance in plants by arbuscular mycorrhizal fungi. Scientia Horticulturae. 2023;**308**

[72] Ren AT, Zhu Y, Chen YL, Ren HX, Li JY, Abbott LK, Xiong YC. Arbuscular mycorrhizal fungus alters root-sourced signal (abscisic acid) for better drought acclimation in *Zea mays* L. seedlings. Environmental and Experimental Botany. 2019;**167**.

[73] Eke P, Adamou S, FokomR N, Vd FPVT, Wakam LN, Nwaga D, Boyom FF. Arbuscular mycorrhizal fungi alter antifungal potential of lemongrass essential oil against Fusarium solani, causing root rot in common bean (Phaseolus vulgaris L.). Heliyon. 2020;**6**(12):e05737

[74] Duan HX, Luo CL, Zhou R, Zhao L, Zhu SG, Chen YL, Zhu Y, Xiong YC. AM fungus promotes wheat grain filling via improving rhizospheric water nutrient availability under drought and low density. Applied Soil Ecology. 2024;**193**

[75] Zhang W, Xia KL, Feng ZW, Qin YQ, Zhou Y, Feng GD, Zhu HH, Yao Q. Tomato plant growth promotion and drought tolerance conferred by three arbuscular mycorrhizal fungi is mediated by lipid metabolism. Plant Physiology Biochemistry. 2024;**208**

[76] Fan XN, Xie HY, Huang XR, Zhang SY, Nie YY, Chen H, XN X, Tang M.A module centered on the transcription factor Msn2 from arbuscular mycorrhizal fungus Rhizophagus irregularis regulates drought stress tolerance in the host plant. New Phytologist. 2023;**240** (4):1497–1518

[77] Diao FW, Jia BB, Wang XH, Luo JQ, Hou YZ, Li FY, Guo W. Proteomic analysis revealed modulations of carbon and nitrogen by arbuscular mycorrhizal fungi associated with the halophyte Suaeda salsa in a moderately saline environment. Land Degradation & Development. 2022;**33** (11):1933–1943

[78] Liu YX, Lu JH, Cui L, Tang ZH, Ci DW, Zou XX ZXJ, Yu XN, Wang YF,

Si T. The multifaceted roles of Arbuscular Mycorrhizal Fungi in peanut responses to salt, drought, and cold stress. BMC Plant Biology. 2023;**23**(1):36

[79] Singh D, Mathimaran N, Boller T, Kahmen A. Bioirrigation: A common mycorrhizal network facilitates the water transfer from deep-rooted pigeon pea to shallow-rooted finger millet under drought. Plant & Soil. 2019;**440** (1-2):277–292

[80] Jia BB, Diao FW, Ding SL, Shi ZQ, Xu J, Hao LJ, Li FY, Guo W. Differential effects of arbuscular mycorrhizal fungi on three salt-tolerant grasses under cadmium and salt stress. Land Degradation & Development. 2023;**34** (2):506–520

[81] Liu SL, Guo XL, Feng G, Maimaitiaili B, Fan JL, He XH. Indigenous arbuscular mycorrhizal fungi can alleviate salt stress and promote growth of cotton and maize in saline fields. Plant & Soil. 2016;**398** (1-2):195–206.

[82] BL W, Wang C, Liu XL, Xue R, Liu ML, Xu GX, Li ZY. Effects of earthworm (Eisenia fetida) and arbuscular mycorrhizal fungi improving plant hormones and antioxidant enzymes under simulated acid rain stress. Applied Soil Ecology. 2023;**182**:104729

[83] Cartmill AD, Valdez-Aguilar LA, Bryan DL, Alarcon A. Arbuscular mycorrhizal fungi enhance tolerance of vinca to high alkalinity in irrigation water. Scientia Horticulturae. 2008;**115** (3):275–284

[84] Khan SR, Ahmad Z, Khan Z, Khan U, Asad M, Shah T. Synergistic effect of silicon and arbuscular mycorrhizal fungi reduces cadmium accumulation by regulating hormonal transduction and lignin accumulation in maize. Chemosphere. 2024;**346**

[85] Madejon E, Doronila AI, Sanchez-Palacios JT, Madejon P, Baker AJM. Arbuscular Mycorrhizal Fungi (AMF) and Biosolids Enhance the Growth of a Native Australian Grass on Sulphidic Gold Mine Tailings. Restoration Ecology. 2010;**18**:175–183

[86] Shen YQ, Wei Y, Chen ZP, Zeng QR, Hou H. Effects of Arbuscular Mycorrhizal Fungi on Growth, Antimony Uptake and Antioxidant Enzymes of Maize under Antimony Stress. Research of Environmental Sciences. 2017;**30**(5):712–719

[87] Wang HR, Du XR, Zhang ZY, Feng FJ, Zhang JM. Rhizosphere interface microbiome reassembly by arbuscular mycorrhizal fungi weakens cadmium migration dynamics. iMeta. 2023;**2**(4)

[88] Biasi R, Brunori E, Vanino S, Bernardini A, Catalani A, Farina R, Bruno A, Chilosi G. Soil–Plant Interaction Mediated by Indigenous AMF in Grafted and Own-Rooted Grapevines under Field Conditions. Agriculture. 2023;**13**(5):1051

[89] Yang YL, Wu ZQ, Liu L, Zhang DH, Ma HC, Wu JR. The relationship between arbuscular mycorrhizal fungi(AMF)and anthracnose occurrence in Camellia oleifera. Journal of Biology. 2023;**40** (5):35–40

[90] Mathur S, Sharma MP, Jajoo A. Improved photosynthetic efficacy of maize (*Zea mays*) plants with arbuscular mycorrhizal fungi (AMF) under high temperature stress. Journal of

Photochemistry & Photobiology, B: Biology. 2018;**180**:149–154

[91] Wang BY, Chen J, Qx C, Bao YM, Wang HN, Qin R, XY LI. Progress of research on culture and application of Arbuscular Mycorrhizal fungi. Bulletin of Botanical Research. 2025;**45**(03): 361–370

[92] Tian XL, Wang CB, Bao XG, Wang P, Li XF, Yang SC, Ding GC, Christie P, Li L. Crop diversity facilitates soil aggregation in relation to soil microbial community composition driven by intercropping. Plant & Soil. 2019;**436**(1):173–192

[93] Xie H, Zhang QF, Chen TT, Zeng CX, Zhou JC, Wu Y, Lin HY, Liu YY, Yin YF, Chen YM. Interaction of soil arbuscular mycorrhizal fungi and plant roots acts on maintaining soil phosphorus availability under nitrogen addition. Chinese Journal of Plant Ecology. 2022;**46**(7):811–822

[94] Zhang L, Feng G, Declerck S. Signal beyond nutrient, fructose, exuded by an arbuscular mycorrhizal fungus triggers phytate mineralization by a phosphate solubilizing bacterium. The ISME Journal. 2018;**12**(10): 2339–2351

[95] Zhang X, Chen B, Ohtomo R. Mycorrhizal effects on growth, P uptake and Cd tolerance of the host plant vary among different AM fungal species. Soil Science and Plant Nutrition. 2015;**61**:359–368

[96] Zhang X, Wu S, Ren B, Chen B. Water management, rice varieties and mycorrhizal inoculation influence arsenic concentration and speciation in rice grains. Mycorrhiza. 2016;**26**: 299–309

[97] Verbruggen E, van der Heijden MGA, Weedon JT, Kowalchuk GA, Roling WF. Community assembly, species richness and nested-ness of arbuscular mycorrhizal fungi in agricultural soils. Molecular Ecology. 2012;**21**:2341–2353

[98] Singh RK, Dai O, Nimasow G. Effect of arbuscular mycorrhizal (AM) inoculation on growth of Chili plant in organic manure amended soil. African Journal of Microbiology Research. 2011;**5**:5004–5012

[99] Labidi S, Ben JF, Tisserant B, Yousfi M, Sanaa M, Dalpe Y, Sahraoui ALH. Field application of mycorrhizal bio-inoculants affects the mineral uptake of a forage legume (Hedysarum coronarium L.) on a highly calcareous soil. Mycorrhiza. 2015;**25**:297–309

[100] Pellegrino E, Turrini A, Gamper HA, Cafa G, Bonari E, Young JPW, Giovannetti M. Establishment, persistence and effectiveness of arbuscular mycorrhizal fungal inoculants in the field revealed using molecular genetic tracing and measurement of yield components. New Phytologist. 2012;**194**:810–822

[101] Azcon R, Rodriguez R, Amora-Lazcano E, Ambrosano E. Uptake and metabolism of nitrate in mycorrhizal plants as affected by water availability and N concentration in soil. European Journal of Soil Science. 2008;**59**:131–138

[102] Masula E, Vassilev N. A contribution to set a legal framework for biofertilisers. Applied Microbiology and Biotechnology. 2014;**98**:6599–6607

[103] Herridge F, Hartley E, Gemell G. Rhizobial counts in peat inoculants vary amongst legume inoculant groups at

manufacture and with storage: Implications for quality standards. Plant & Soil. 2014;**380**(1-2):327–336

[104] Duan HX, Luo CL, Li JY, Wang BZ, Naseer M, Xiong YC. Improvement of wheat productivity and soil quality by arbuscular mycorrhizal fungi is density-and moisture-dependent. Agronomy for Sustainable Development. 2021;**41**(1):3

[105] Johnson NC, Angelard C, Sanders IR, Kiers ET. Predicting community and ecosystem outcomes of mycorrhizal responses to global change. Ecology Letters. 2013;**16**:140–153

[106] Gou XM, Hu YX, Ni HQ, Wang X, Qiu LP, Chang XC, Shao MG, GH W, XR W. Arbuscular mycorrhizal fungi alleviate erosional soil nitrogen loss by regulating nitrogen cycling genes and enzymes in experimental agroecosystems. Science of the Total Environment. 2024;**906**

[107] Yu H, Liu XY, Yang C, Peng YS, Yu XL, Gu H, Zheng XF, Wang C, Xiao FS, Shu LF, He ZL, Wu B, Yan Q. Co-symbiosis of arbuscular mycorrhizal fungi (AMF) and diazotrophs promote biological nitrogen fixation in mangrove ecosystems. Soil Biology Biochemistry. 2021;**161**

[108] Bender SF, Plantenga F, Neftel A, Jocher M, Oberholzer HR, Kohl L, Giles M, Daniell TJ, der Heijden Mga V. Symbiotic relationships between soil fungi and plants reduce N2O emissions from soil. The ISME Journal. 2014;**8**(6):1336–1345

[109] Leigh J, Hodge A, Fitter AH. Arbuscular mycorrhizal fungi can transfer substantial amounts of nitrogen to their host plant from organic material. New Phytologist. 2009;**181**(1):199–207

[110] Vosátka M, Latr A, Gianinazzi S, Albrechtova J. Development of arbuscular mycorrhizal biotechnology and industry: Current achievements and bottlenecks. Symbiosis. 2012;**58**:29–37

[111] Antunes PM, Koch AM, Dunfield KE, Hart MM, Downing A, Rillig MC, Klironomos JN. Influence of commercial inoculation with Glomus intraradices on the structure and functioning of an AM fungal community from an agricultural site. Plant & Soil. 2009;**317**:257–266

[112] Meddad A. Effects of arbuscular mycorrhizal fungi and fertilization levels on industrial tomato growth and production. International Journal of Agriculture and Biology. 2017;**19**:341–347.

[113] Xie W, Hao Z, Zhou X, Jiang X, Xu L, Wu S, Zhao A, Zhang X, Chen B. Arbuscular mycorrhiza facilitates the accumulation of glycyrrhizin and liquiritin in Glycyrrhiza uralensis under drought stress. Mycorrhiza. 2018;**28**:285–300

[114] Bi YL, Li XL, Wang HG. Christie P. Establishment of dual culture between vesicular-arbuscular mycorrhizal fungus Sclerocystis sinuosa and transformed Ri T-DNA carrot roots in vitro. Plant & Soil. 2004; **261**:239–243

[115] Chen BD, Li XL, Zhu YG. Characters of metal adsorption by AM fungal mycelium. Mycosystema. 2005;**24**(2):283–291

Section 3

Biotechnology and Bioinformatics in Mycorrhizal Fungi Research

Chapter 4

Advanced Bioinformatics Explores Mycological Composition and Function from Metatranscriptomic Sequencing

Haihua Wang

Abstract

Fungi constitute an immensely diverse component in ecosystems, driving key ecosystem processes such as decomposition, nutrient cycling, and host interactions. Metatranscriptomic sequencing provides the chance to look into this activity by capturing the pool of RNA transcripts produced *in situ*, thereby linking taxonomic identity with functional expression. By recovering ribosomal RNA (rRNA) or single-copy marker genes alongside messenger RNA (mRNA), metatranscriptomes enable integrated "who is active and what they are doing" profiles that overcome many limitations of marker-gene associated amplicon sequencing. Typical workflows extract total environmental RNA, deplete abundant rRNA, convert remaining RNA to complementary DNA (cDNA), and then sequence; downstream bioinformatic pipelines partition rRNA for taxonomic profiling and mRNA for functional annotation, using either *de novo* assembly or direct mapping to reconstruct transcripts and infer enzymatic activities. Major challenges remain, however, complex eukaryotic transcript structures, low relative abundance of fungal RNA amid host and bacterial backgrounds, and incomplete, biased fungal reference genomes and functional databases limit taxonomic resolution and annotation rates. Continued advances, including long-read RNA sequencing, expanded fungal genome initiatives, and machine learning–driven functional prediction, promise to enhance sensitivity and interpretability. By directly coupling fungal identity with gene expression, metatranscriptomics is poised to substantially advance our mechanistic understanding of fungal ecology and ecosystem function. In this chapter, we will summarize the research advances in the bioinformatic technologies used for mycological taxonomic and functional classification from metatranscriptomic sequencing datasets.

Keywords: metatranscriptomics, mycobiome, taxonomic classification, function profiling, bioinformatics pipelines, environmental ecology

1. Introduction

Fungi are diverse and widespread organisms on earth, with an estimated 2.2–3.8 million species inhabiting virtually all ecosystems [1]. Global studies consistently show the highest fungal diversity in soils, followed by freshwater and marine habitats (see **Figure 1**) [2]. In terrestrial and freshwater environments, fungi dominate the decomposition of organic matter and mediate major element cycles and mineralizing nitrogen and phosphorus. They also engage in symbiotic relationships (e.g., mycorrhizae and endophytes) and act as pathogens or mutualists, playing central roles in nutrient cycling, carbon storage, and host health across forests, grasslands, aquatic systems, and beyond [3]. These ecological functions make it critical to accurately characterize fungal communities and their activities in environmental samples.

Traditionally, fungal community studies have relied on amplicon sequencing of ribosomal markers, such as the ITS or 18S ribosomal RNA (rRNA) regions [4]. However, primer mismatches and variable PCR amplification can introduce large biases in which taxa are detected [5]. In addition, fungal ribosomal gene copy numbers vary dramatically between species [6], resulting in a bias in fungal abundance estimation. In short, while rapid and low-cost, the ITS/18S amplicon approach provides only a limited and potentially distorted view of mycobiome

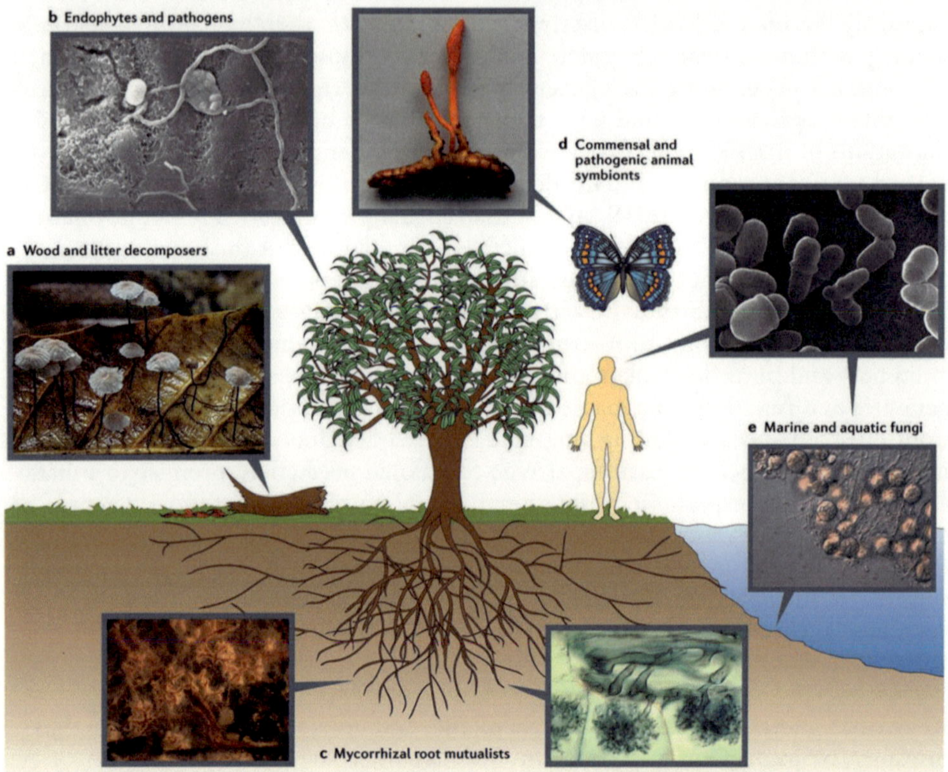

Figure 1.
Functional diversity of the mycobiome. Fungi are most commonly associated with terrestrial ecosystems, but can also be found growing on nearly any substrate on earth, from deep ocean sediments to the human scalp [74].

composition, metatranscriptomic sequencing of total RNA offers an alternative that overcomes many of these limitations (see **Figure 2**) [7]. By RNA sequencing, metatranscriptomics directly captures the pool of actively transcribed genes *in situ* (see **Figure 3**). Each metatranscriptome thus provides a profile of the expressed genetic functions and taxa present, enabling both taxonomic and functional profiling. For example, metatranscriptomics can recover rRNA or single-copy marker genes for taxonomic identification and messenger RNA (mRNA) transcripts for functional genes and pathways. Meanwhile, metatranscriptomics is less prone to primer bias and can reveal which fungal genes (e.g., carbohydrate-active enzymes [CAZymes], transporters, secondary metabolites) are being expressed under natural

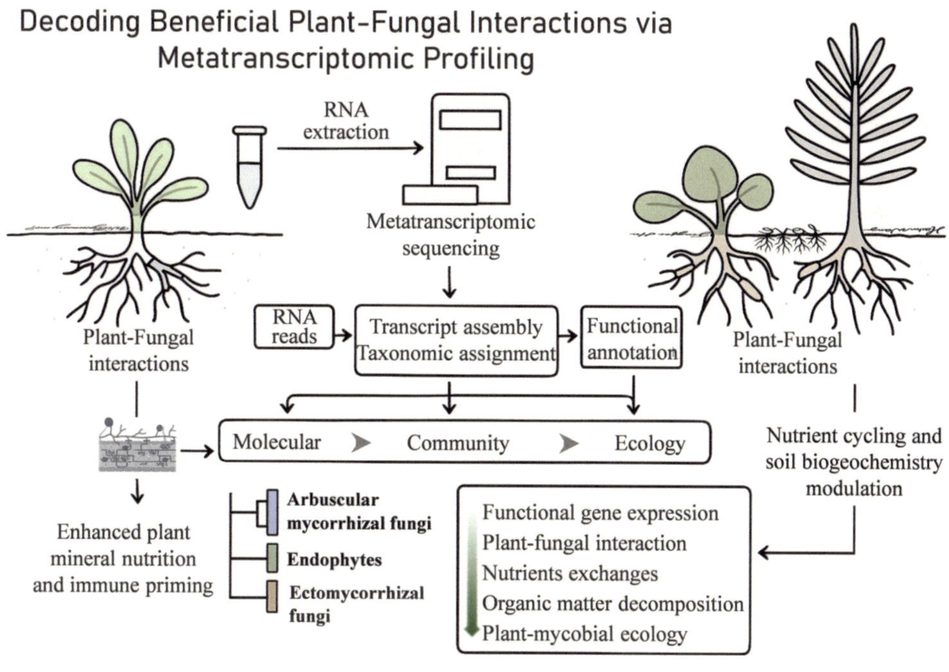

Figure 2.
Schematic workflow for decoding beneficial plant–fungal molecular interactions using metatranscriptomic profiling. This diagram is conceptual and illustrates how combined taxonomic and functional sequencing enables a "who is active and what they are doing" view of plant-associated mycobiomes.

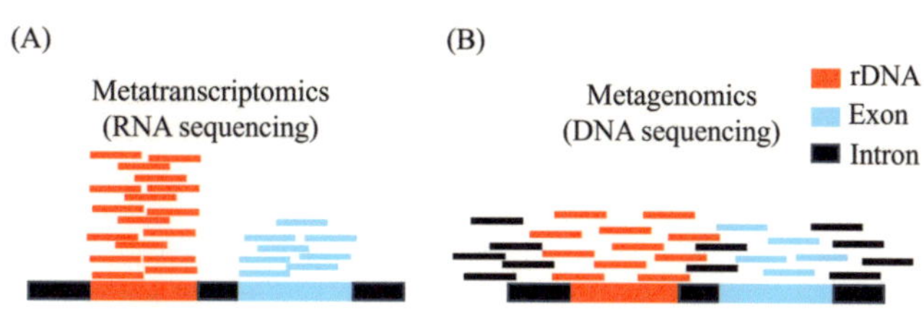

Figure 3.
Conceptual charts illustrate metatranscriptomic and metagenomic sequencing. Metatranscriptomics targets the functional, expressed portion of the fungal (or microbial) community by profiling active RNA transcripts, whereas metagenomics provides a broader snapshot of the entire genomic potential present in the sample, including non-expressed regions.

conditions [8]. In practice, metatranscriptomics enables us to have an insight into microbial community activity and interactions, such as which members of the mycobiome are metabolically active and what ecological processes (decomposition, nutrient uptake, stress responses, etc.) are occurring.

In a typical metatranscriptomic workflow (see **Figure 4**) [9], total RNA is extracted from an environmental sample (soil, leaf litter, water, etc.), then rRNA is depleted, and the remaining RNA is reverse-transcribed and sequenced on a sequencing platform. The raw reads are quality-trimmed and filtered for contaminants (adapters, host DNA, etc.). Pipelines then separate rRNA (for taxonomic profiling) and mRNA (for functional classification) reads. For example, MetaPro emphasizes filtering out low-quality and host/rRNA reads, then assembles

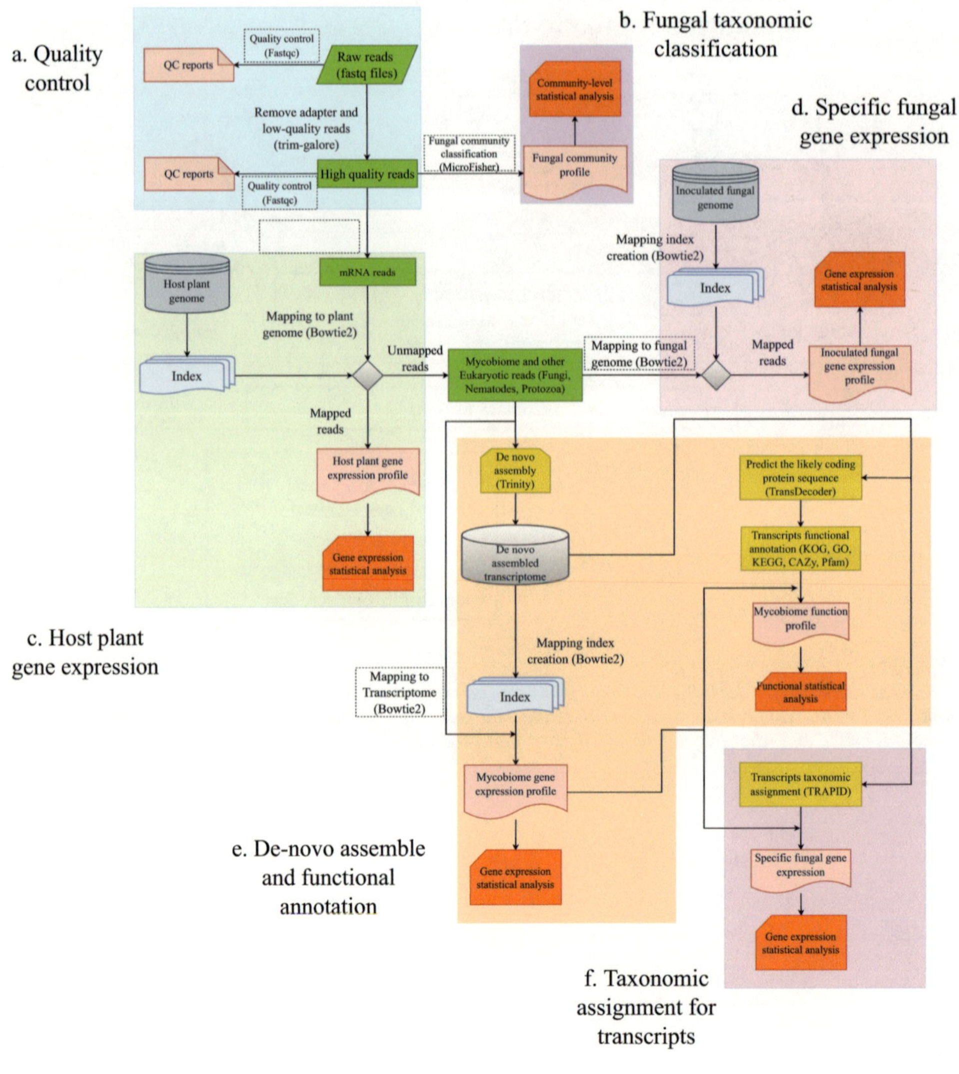

Figure 4.
Pipeline for fungal metatranscriptomic data processing and interpretation. The diagram is organized into six labeled modules that outline a typical end-to-end workflow from raw reads to taxon-resolved functional inference.

transcripts and annotates them to produce gene expression tables [10]. MetaTrans similarly removes rRNA reads and aligns the non-rRNA reads to functional gene databases to quantify expression levels [11]. FunOMIC aligns shotgun reads directly to its fungal marker gene and protein databases to assign fungal species and functions [12]. In all cases, the output includes lists of fungal genes and enzymes with expression abundances, linked to taxonomic identities. Downstream analysis can then reconstruct active pathways, enzyme networks, or diversity metrics from these data. Hesse et al. [13] sequenced forest-floor metatranscriptomes from northern hardwood forests and found that long-term nitrogen deposition altered fungal (and bacterial) gene expression in litter. In that study, they simultaneously profiled carbohydrate-active enzyme transcripts and observed significant changes in CAZyme expression under nitrogen fertilization, even when operational taxonomic unit (OTU) richness remained unchanged. In another example, Zeng et al. [14] analyzed soil metatranscriptomes from old-growth montane forests. They observed that, despite large species turnover between forests, the functional gene expression profiles (e.g., proteases, metabolic enzymes related to decomposition and N acquisition) were remarkably stable across sites. This "stable functional structure" suggests that core ecological functions are maintained by different fungal taxa in each forest. In a controlled experiment, Marcelino et al. [15] demonstrated that metatranscriptomic sequencing of mixed fungal communities could recover species- and strain-level taxa with high accuracy, whereas single-marker metabarcoding had insufficient phylogenetic resolution. Collectively, these studies show that metatranscriptomics can simultaneously reveal which fungal species are active in an environment and what metabolic processes they are performing, providing insights that greatly extend beyond what amplicon surveys alone can provide.

2. Challenges in fungal metatranscriptomic analysis

Fungal metatranscriptomics has emerged as a powerful tool to link community composition with function by directly capturing the RNA of active fungi *in situ*. Unlike DNA studies, RNA sequencing profiles the functional gene expression, revealing dynamic metabolic activity and pathway utilization. Recent studies illustrate that metatranscriptomics can uncover upregulated enzyme genes and coherent shifts in metabolic pathways corresponding to environmental changes, thus identifying the functional roles of diverse fungal taxa (see **Figure 2**) [8, 16, 17]. Despite these strengths, however, fungal systems pose significant challenges. Fungal transcriptomes are highly complex and heterogeneous, especially in environmental samples, making assembly and mapping difficult. In many host–fungus or environmental samples, host or other non-target RNA dominates the library, so that fungal mRNAs are under-represented. In addition, current fungal reference databases are far from complete, many fungal genomes and transcripts are missing or poorly annotated, which limits taxonomic resolution and functional assignment of RNA reads.

2.1 High complexity of fungal transcriptomes and mRNA contamination

Fungal transcriptomes are inherently complex and pose unique challenges. Many fungi have large and repetitive genomes, for example, the rust fungi (*Pucciniales*) have an average of 300 Mb and range up to 900 Mb, which often contain many duplicate genes, transposons and introns [18–20]. Fungal genes are interrupted by introns and undergo alternative splicing (AS). Indeed, widespread AS has been documented in pathogenic fungi, one study in the rice blast fungus *Magnaporthe oryzae* found 4,270 AS isoforms from 2,413 genes, with a third of these isoforms expressed only during host infection [21, 22]. Many of the dominant isoforms encoded regulatory proteins, implying functional diversification. These features complicate transcriptome assembly, as assemblers must identify intron–exon boundaries and discriminate closely related isoforms and paralogs. The popular metagenome assemblers (designed for prokaryotes) assume genes are contiguous, as noted, and are ineffective in microbiomes that also contain eukaryotes because they cannot handle splicing. In addition, fungal cells can vary in ploidy or chromosome copy number (aneuploidy), especially under stress or during infection. Such copy-number variation can bias expression measurements. Finally, fungal gene expression is highly dynamic (e.g., stage-specific infection genes or environmental responses), which adds another layer of complexity [23–25]. Together, large genomes with many introns and paralogs, extensive AS, and variable ploidy make accurate *de novo* assembly and quantification of fungal transcripts much more difficult than for typical bacteria or simple eukaryotes.

In environmental or host-associated samples (see **Figure 2**), extracting fungal RNA without contamination is technically challenging because fungal RNA is often a tiny fraction of total RNA. In host tissues (plant or animal), host RNA far outnumbers fungal RNA, making direct sequencing of fungal transcripts inefficient. Likewise, in soil or plant environments, co-extracted substances (e.g., humic acids, phenolic compounds) and ubiquitous RNases degrade RNA and inhibit enzymes. These inhibitors and background RNases must be removed by a rigorous extraction pipeline to recover high-quality RNA. Finally, cross-domain contamination can confound analysis; bacterial, plant, and viral sequences are frequently co-extracted and sometimes misattributed as fungal reads. In practice, stringent bioinformatic filtering is required, but incomplete host references can still lead to misclassification. Overall, separating fungal RNA from plant, animal, or bacterial RNA remains a challenge in the analysis of fungal metatranscriptomics.

2.2 Incomplete and biased reference databases

A recent study demonstrated that capturing the mycobiome taxonomic composition from metagenomic data shows a significant lack of comprehensive databases [26]. Current fungal reference databases are heavily skewed toward a few model groups and lack coverage of most fungal diversity. Most sequenced fungal genomes are from the Dikarya (Ascomycota and Basidiomycota), while early-diverging lineages (e.g., many zygomycetes and basal clades) remain under-represented [27, 28]. Even where sequences exist, many assemblies are incomplete drafts or unavailable. As a result, standard reference sets only collect genes and pathways of specific fungal groups that have genome reference. In practice, this means that many environmental fungal transcripts have no close match in the database, hampering taxonomic assignment and functional prediction in the metatranscriptomic analysis. In summary, the paucity and unevenness of fungal reference genomes and transcriptomes, coupled with non-standardized or shallow annotations, severely limit the sensitivity and accuracy of downstream analyses.

Moreover, functional annotation of fungal metatranscriptomes is constrained by the lack of databases. Currently, annotation pipelines typically use broad databases (KEGG, COG, Pfam) that were built on bacterial or model eukaryote data [29–31], hence, many fungal pathways are missing or poorly characterized. This leads to low annotation rates for fungal-specific genes. A study of *Grifola frondosa* found that many secondary metabolism genes were mis-annotated or unrecognized because the database lacked close homologs [32]. Novel transcripts, especially those from unexplored clades, often have no significant BLAST (Basic Local Alignment Search Tool) hits at all. Even the KEGG Orthology (KO) assignments tend to map divergent fungal genes to distant homologs, yielding misleading pathway enrichment results. In short, most annotated functions reflect only a subset of "housekeeping" metabolism (glycolysis, translation, etc.), while fungal-specific enzymes, such as lignocellulose breakdown, unique secondary metabolites, and effector proteins, lack coverage. Until fungal genomes are better annotated, and until specialized resources grow, many fungal transcripts will remain uncharacterized. This forces reliance on *de novo* function prediction, which still misses many candidate genes.

2.3 Lack of standardized pipelines for fungal metatranscriptomics

Unlike amplicon analysis, there is no consensus pipeline for metatranscriptomic sequencing dataset. Most studies have used a heterogeneous mix of tools for quality control, rRNA removal, assembly, and annotation. Only very few workflows are tailored to fungi or benchmarked on mixed communities. For example, the popularly used metatranscriptome tools (e.g., metaSPAdes, MEGAHIT) were originally developed for prokaryotes or single eukaryotes [33, 34], and their performance on mixed fungal data is uncertain, whereas only a handful of methods have been specifically developed for metatranscriptomics at all, and their efficacy is largely untested. Similarly, the mycobiome community evaluation found only a few taxonomic classifiers (MiCop, FunOMIC, etc.) that even include fungi, and none agreed fully on community composition [12, 35]. The result is poor reproducibility; analyzing the same sample can yield different taxa and functional profiles simply due to pipeline choice. Inconsistent choices of databases and parameters make it hard to compare studies, hindering meta-analysis and broad conclusions.

Moreover, fungal genomes are under-represented in reference databases, current taxonomic classifiers often miss or mis-identify fungi from metatranscriptomic datasets. Tools like Kraken2 [36, 37, 38], Centrifuge, or MetaPhlAn rely on aligning reads against known references or genomes that work well when the target organism has close relatives in the database. But for novel or rare fungi, these read-based methods often return "no hits" or low-confidence assignments. Benchmarks have shown that when multiple related species exist, classifiers frequently detect only one representative per genus, even if other species are present. It means those pipelines perform well in fungal classification at genus and higher taxonomic levels. Many classifiers also have limited fungal reference sets, since those tools do not have eukaryotes as part of their databases. Thus, low-abundance fungal reads may lack unique targets and either remain unclassified or be falsely assigned to the wrong taxon. Collectively, the overall result is low sensitivity for fungi at the species level and high error rates for closely related strains, especially when relying on short or poor-quality reads.

2.4 Future directions for overcoming challenges

Several emerging approaches promise to mitigate these limitations. Long-read sequencing (e.g., PacBio Iso-Seq or Oxford Nanopore) can capture full-length fungal transcripts [39–41], resolving splice variants and repetitive regions that short reads miss. Recent studies show that nanopore sequencing of extended ITS/LSU amplicons can achieve near-species-level resolution for fungi, suggesting similar gains for transcripts. At the same time, reference resources are expanding, initiatives like the DOE JGI 1000 Fungal Genomes project [42] aim to sequence thousands of new genomes across the tree of life, and databases like FungiDB [43] are continuously updated. Machine-learning methods (e.g., deep learning annotation tools) are also being developed to predict gene function from sequence features, which could help assign tentative roles to novel fungal genes. Finally, community efforts towards standardization, such as containerized workflows, open data sharing, and benchmarks, are beginning to emerge. By combining longer reads, richer reference databases, improved annotation algorithms, and shared best practices, future studies should greatly improve the sensitivity, accuracy, and reproducibility of fungal metatranscriptomics.

3. Bioinformatic tools for mycological taxonomic classification

Classifying fungal taxa from metatranscriptomic data typically follows two main bioinformatic strategies: assembly-based reconstruction of transcripts and alignment-based read-mapping to references (see **Figure 5**). In the assembly-based approach, quality-filtered RNA-Seq reads are assembled *de novo* into longer contigs or transcripts, which are then annotated for taxonomic or functional assignment. This strategy provides longer sequence scaffolds, making it easier to find homologous genes and assign taxonomic origin [44, 45]. In contrast, alignment-based methods map reads directly to reference sequences using k-mers, protein alignment, or marker-gene matching. Both approaches have been applied to fungal community profiling, and each has strengths and limitations in handling fungal genome complexity, reference coverage, and mixed-community noise [46].

3.1 Assembly-based taxonomic classification

De novo assembly of metatranscriptomic data aims to reconstruct contiguous marker genes or transcripts from reads, followed by a taxonomic assignment through mapping those transcripts onto reference databases. Longer contigs can be more specifically and accurately matched to reference databases, enabling more confident taxonomic assignment and the discovery of novel sequences. Assembled transcripts provide a reference scaffold representing longer expressed genome segments that facilitate homology searches and taxonomic assignment [45]. In practice, assembly

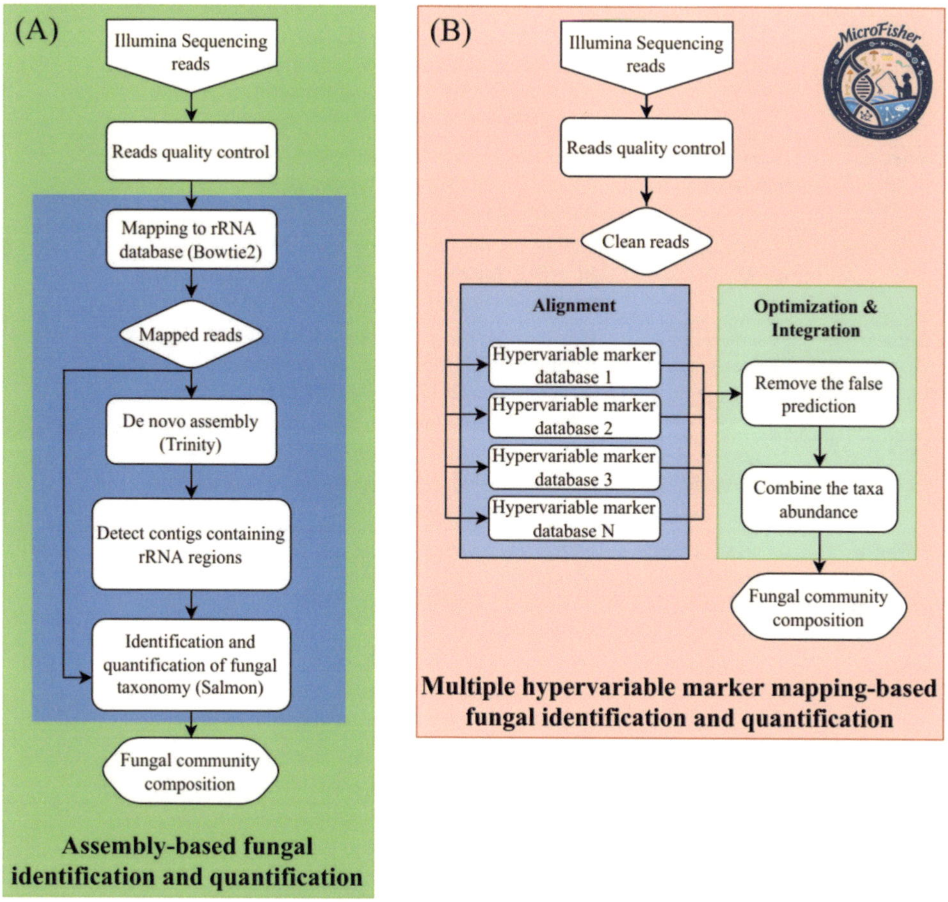

Figure 5.
Comparison of two pipelines for recovering fungal community composition from illumina metatranscriptomic reads. The figure shows two panels: (a) Assembly-based fungal identification and quantification and (b) multiple hypervariable-marker mapping-based fungal identification and quantification.

can yield near full-length fungal transcripts that often contain several hundred to thousands of base pairs, improving resolution compared to short reads alone. Common tools for transcriptome assembly include Trinity [47], Oases [48], and rnaSPAdes [49], which are designed for RNA-seq; other tools like Trans-ABySS [50], SOAPdenovo-Trans [51] and BinPacker [52] are also used for *de novo* assembly of metatranscriptomic data (see **Table 1**). Previous study found that assembly with Trinity or Oases yielded more annotated transcripts than general assemblers, with Trinity giving the best recovery of genes [53]. This suggests that assemblers tailored to transcript data often outperform general metagenomic assemblers in recovering fungal genes. These assemblers produce contigs that are then annotated by BLASTn searches onto NCBI nt database followed by taxonomic binning or directly matching with the UNITE and RDP databases for classification. Meanwhile, using mRNA transcript annotation results from eggNOG-mapper enables taxonomic assignment with TRAPID [54].

Tool	Function	Features	Reference
Trinity	*De novo* transcriptome assembler	Reconstructs full-length transcripts without a reference genome. Recovers more complete transcripts across expression levels than many other assemblers.	[75]
Oases	*De novo* transcriptome assembler	Heuristically assembles RNA-seq reads which could handle alternative splicing. Improves on Trinity and Trans-ABySS in benchmarks.	[48]
rnaSPAdes	*De novo* transcriptome assembler	Built on the SPAdes genome assembler, optimized for RNA-Seq. Typically yields more assembled genes and isoforms and higher accuracy than most competitors.	[49]
Trans-ABySS	*De novo* transcriptome assembler	Extends ABySS assembler for RNA-Seq which assembles reads at multiple k-mer sizes and merges contigs. Handles variable coverage; high sensitivity and specificity in mouse transcriptome.	[50]
SOAPdenovo-Trans	*De novo* transcriptome assembler	Adapts SOAPdenovo for RNA data; designed for short reads. Produces high-contiguity transcripts with low redundancy and runs quickly.	[51]
BinPacker	*De novo* transcriptome assembler	Novel "bin-packing" approach to assemble isoforms. Incorporates coverage information to resolve isoforms; outperforms most other assemblers in tests, with low memory use.	[52]
MEGAHIT	Metagenome assembler	Ultra-fast *de novo* assembler for large (meta)genomic datasets. Not specific to transcripts (no splice handling), but sometimes used for mixed microbial assemblies.	[33]
DIAMOND	Sequence aligner	Ultra-fast protein aligner. Provides high sensitivity alignment, suitable for mapping transcripts to nr.	[76]
BLASTX	Sequence aligner	Standard NCBI tool for translating nucleotide queries and aligning to protein databases. Widely used for transcript annotation via similarity searches.	[77]
BASTA	Taxonomy classifier	Command-line lowest common ancestor (LCA) classifier. Takes BLAST/DIAMOND outputs and assigns sequences to taxa via customizable LCA rules.	[78]
TRAPID	Transcript taxonomic assignment tool	Online system for *de novo* transcriptomes, ORF prediction, frameshift correction, gene family assignment, and functional and phylogenetic analysis.	[54]

Table 1.
Comparison of bioinformatic tools for assembly and taxonomic classification of fungal mRNA transcripts.

The advantages of assembly-based classification include increased specificity and discovery of novel taxa. Longer transcripts allow alignment to unique regions (e.g., coding genes beyond ribosomal markers) and can span variable domains that distinguish closely related species. This approach also reconstructs expressed genes for downstream functional analysis. For instance, Schneider et al. [8] developed an assembly pipeline for spruce root samples, they assembled 615,331 fungal transcripts and were able to assign taxonomies to 95.5% at the phylum level and 34.2% at the species level. Importantly, assembly can reveal fungi not in standard databases; if a contig has no match, it may represent a novel lineage, prompting further investigation. However, *de novo* assembly requires significant CPU time and RAM usage, especially for complex samples with many organisms. In addition, low-abundance transcripts may fail to assemble, biasing results toward dominant fungi [55]. Chimeric contigs can occur when similar transcripts from different strains or species are assembled into one sequence. Consequently, resolving very closely related fungi may be challenging. Strain-level variants could be distinguished only when transcripts were highly abundant, and very low-abundance detections (<0.5% of reads) often represented contaminants or errors. In practice, users often discard taxa supported by extremely few contigs to avoid false positives. Furthermore, assemblers originally designed for genomes may struggle with eukaryotic splicing; Shakya et al. [45] note that metagenomic assemblers like MEGAHIT have not been fully validated on spliced transcripts and may misassemble isoforms.

In summary, assembly-based classification offers improved resolution by reconstructing full transcripts, enabling both taxonomy and functional insight from mycobiomes. It outperforms at specificity, thereby discovering novel or divergent fungi. Assembly-based profiling could accurately recover most species in a controlled mix, but recommends filtering out hits below 0.5% abundance to eliminate spurious assignments. Similarly, Celaj et al. [56] found that assemblies from Trinity yielded more annotated genes than other methods on mixed data, highlighting the choice of assembler matters. Schneider et al. [8] reported that the assembly-based pipeline annotated most fungal transcripts in forest samples, but only one-third reached species-level resolution. These studies suggest that assembly is effective for revealing community composition when sequencing depth and computational resources are sufficient, but it may miss rare taxa and requires careful validation. However, this approach needs higher computational cost, potential assembly errors, and difficulty resolving closely related or low-abundance members. Thus, this strategy is most suitable when deep sequencing is available, and one seeks comprehensive, reference-independent profiling of the fungal transcriptome.

3.2 Alignment-based taxonomic classification

Alignment-based methods classify fungal taxa by directly mapping sequencing reads to existing databases. These approaches avoid assembly and thus run much faster with lower memory requirements. They rely on either exact k-mer matching, index-based read alignment, or targeted marker gene alignment. In practice, one aligns raw or preprocessed reads (or predicted genes) to fungal reference genomes, proteins, or marker libraries (see **Table 2**). Alignment-based methods are typically fast and scalable. For example, Kraken2 can classify tens of millions of reads in minutes on a standard server [57]. Centrifuge was explicitly designed to run on limited hardware [37]. These methods bypass the heavy computation of assembly, making them ideal for large multi-sample datasets or initial screening. They also avoid assembly artifacts without chimeras and can be applied incrementally as new data arrive. In practice, classification of metatranscriptomic reads can be done on large RNA-Seq projects with these tools.

However, alignment-based tools are limited by reference completeness and read length throughput that are highly scalable and amenable to large-scale studies [26]. Short reads carry limited taxonomic fingerprint information, so classifiers often only assign higher taxonomic ranks unless the species is well represented in the reference. Kraken2 and Centrifuge match reads to databases generated from genomes, hence they will miss any fungal species whose genome is absent from the database. Similarly, MetaPhlAn4 [38] only detects organisms that have unique marker sequences defined, so

Tool	Category	Algorithmic approach	Database	Strength	Limitation	References
Kraken2	Genome-based	Exact *k*-mer matching with compact index	Genome database	Extremely fast; and scalable	Sensitive to database completeness and limited fungal coverage	[57]
Centrifuge	Genome-based	FM-index on concatenated sequences	Genome database	Low memory use and fast	Limited fungal resolution and false positives	[37]
HumanMycobiomeScan	Genome-based	BLAST-based genome mapping	Fungal genomes	Straightforward alignment and human focus	Computationally heavy	[79]
FindFungi	Genome-based	Custom *k*-mer libraries	Fungal genomes	Fungal-specific tool	Small database	[80]
MEGAN	Genome-based	BLAST/DIAMOND plus LCA assignment	NCBI or custom database	Broad taxonomic profiling and visualization support	Requires pre-alignment; slower on big datasets	[81]
CCMetagen	Genome-based	Custom alignment plus LCA	Curated microbial database	Accurate fungal IDs and precision classification	Less adopted	[82]
MiCoP	Genome-based	Read alignment plus LCA	Genome database	Strong fungal detection; benchmarked	Reference-dependent	[35]
Taxator-tk	Genome-based	Phylogenetic neighborhood inference	Genome DB	Evolutionary-aware taxonomy	Computationally heavy and not fungal-specific	[83]
Kaiju	Protein-based	Translates reads to proteins and aligns to database	Protein DB	Detects distant homologs (proteins conserved)	Coding seqence only and heavy compute	[58]
FunOMIC-P	Protein-based	Protein alignment for taxonomy and function	Protein database	Functional and taxonomic profiling	Mainly human fungi; limited environmental coverage	[12]
VEBA	Hybrid (Genome and Protein)	Integrates genome and protein taxonomic and functional annotation	Cross-domain database	Cross-domain (fungi, bacteria, viruses)	Complex and resource-intensive	[84]
MetaPhlAn4	Marker-based	Bowtie2 alignment to clade markers	Marker database	Species-level abundances	Limited to taxa with defined markers	[38]
mOTUs2	Marker-based	Universal single-copy markers	Prokaryotic DB	Robust for bacteria and scalable	Minimal fungal coverage	[85]

DOI: http://dx.doi.org/10.5772/intechopen.1012591

Tool	Category	Algorithmic approach	Database	Strength	Limitation	References
EukDetect	Marker-based	Conserved eukaryotic protein markers	Marker genes from eukaryotic genomes	Sensitive to microbial eukaryotes	Misses novel fungi	[60]
MicroFisher	Marker-based	Hypervariable ribosomal markers (ITS1, ITS2, LSU D1/D2)	Curated fungal marker database	High sensitivity and accurate	Very new; database expanding	[59]
FunOMIC-T	Marker-based	Bowtie2 alignment to fungal marker genes	Fungal genomes	Fungal-focused	Human-centric database and incomplete coverage	[12]
EUKulele	Marker-based (microeukaryotes)	Taxonomic annotation of microeukaryotes	Curated database	Sensitive to environmental fungi and eukaryote-aware	Less tested	[86]

Table 2.
Comparison of alignment-based bioinformatic tools for fungal taxonomic classification.

uncharacterized fungi will be invisible. Protein-based Kaiju [58] can catch more distant homologs, but it cannot classify non-coding sequences. A challenge is that fungal genome reference databases are far from complete; environmental fungi (e.g., soil saprotrophs, plant endophytes) may have no close genome available, so they won't be detected. In addition, misclassification can occur if a read matches a conserved region shared by multiple taxa, most tools then report a lowest common ancestor (LCA), which may be a genus or family rather than species. Thus, MicroFisher [59] generated hypervariable marker databases, using Internal transcribed spacer (ITS) and large-subunit (LSU) rRNA, by identifying and extracting the highly diverse gene regions from millions of fungi (see **Figure 6**), providing a standardized approach to classify and normalize fungal composition from metatranscriptomic datasets (see **Figure 5b**). Wang et al. [59] indicated that combining classification results using multiple marker genes with weight-based algorithms outperforms simply using single marker classification alone (see **Figure 7**). In addition, incomplete taxonomic coverage also affects marker-based profilers. For instance, the fungal component in the MetaPhlAn4 tool is growing, but still covers a small fraction of known mycobiome diversity. In practice, alignment-based classifiers often report only well-known or abundant taxa, and novel or rare fungi might be missed or only identified at high taxonomic levels.

Benchmarking these tools specifically on fungal metatranscriptomes is limited, but related studies shed light. Avershina et al. [26] tested several tools on synthetic mycobiomes (see **Figure 8**). They found that performance varied greatly, Kraken2

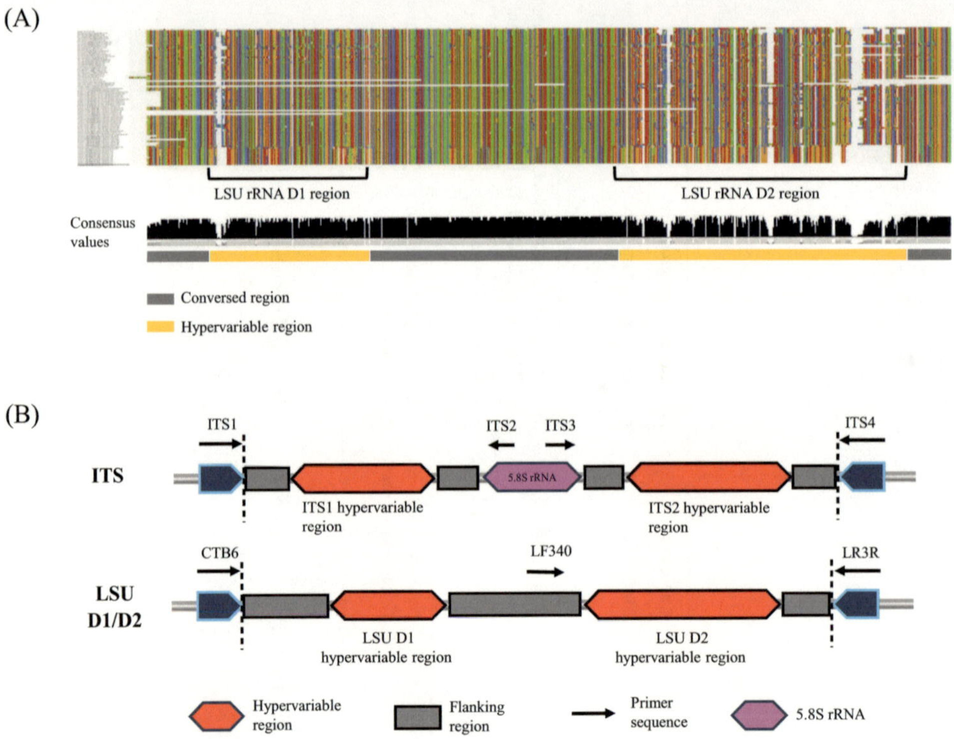

Figure 6.
(a) Representative sequence alignment example showing the sequence polymorphisms of LSU D1 and LSU D2 regions and the hypervariable regions. (b) Conceptual plots illustrate the structure and positions of hypervariable regions in ITS (above) and LSU rRNA D1/D2 (below) sequences.

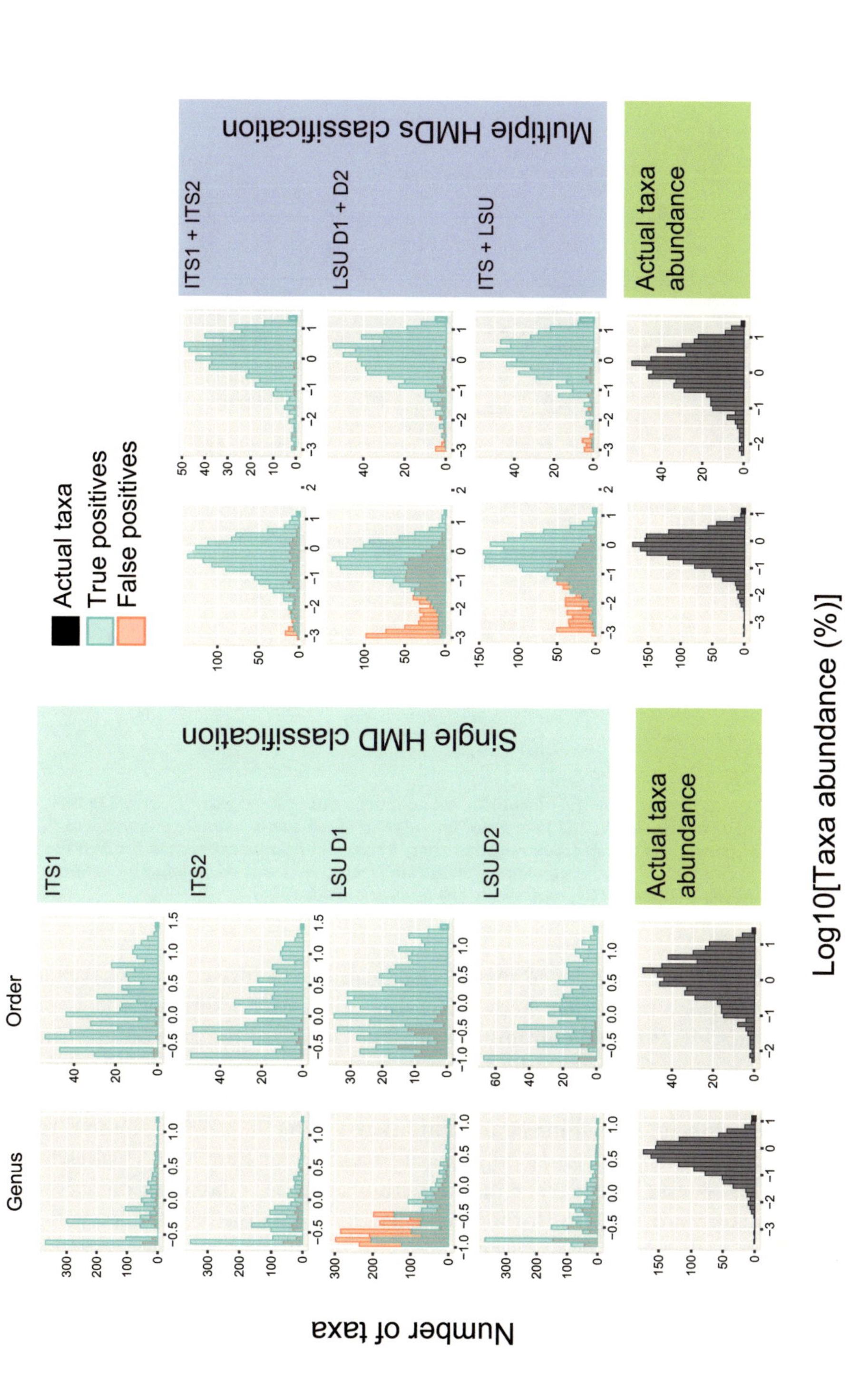

Figure 7.
Histogram plots compare the patterns of the fungal abundance (genus and order levels) between predicted results and actual data using single and multiple marker genes [59].

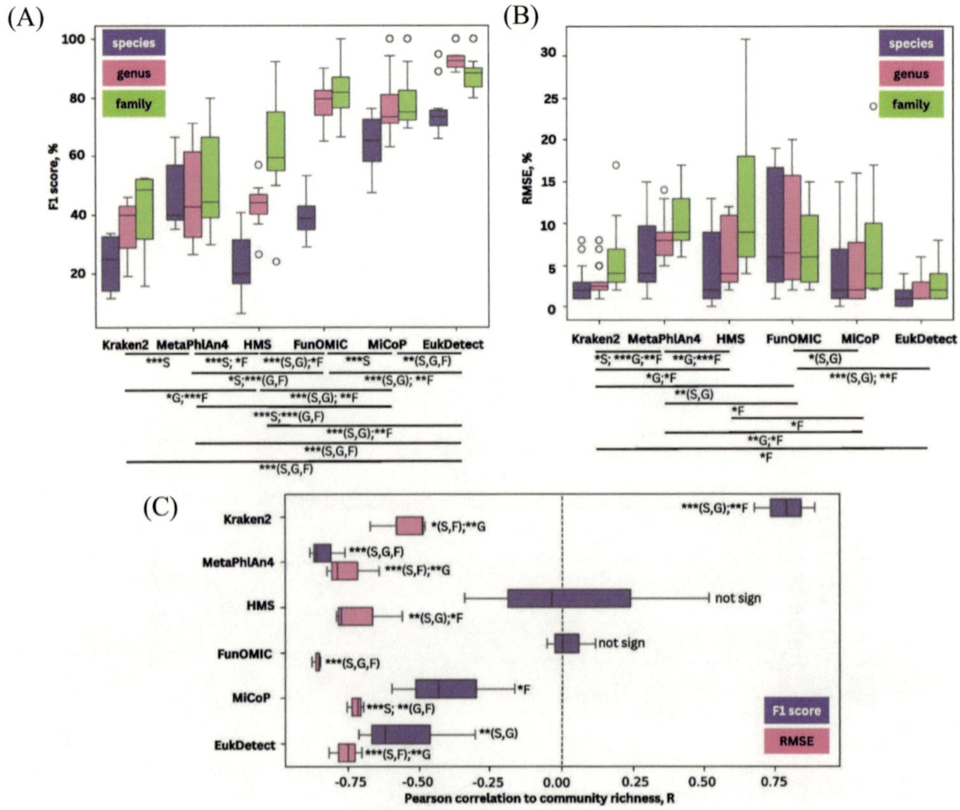

Figure 8.
Accuracy of Fungi characterization. (A) Identification accuracy at the species (purple), genus (green), and family (dark red) taxonomy levels. (B) Root mean square error of relative abundance prediction at the species (purple), genus (green), and family (dark red) taxonomy levels. (C) Pearson correlation between the identification accuracy (purple) or relative abundance prediction error (pink) and the community richness. Significance at species (S), genus (G), and family (F) levels is provided [26].

and MetaPhlAn4 often produced many false positives or missed species, whereas specialized profilers performed better. Notably, FunOMIC identified most species, and EukDetect [60] yielded community profiles closest to the true composition. In that study, only one fungus (*Candida orthopsilosis*) was consistently detected by all tools across all communities, emphasizing that no single method is universally reliable. Importantly, Avershina et al. found that adding 90%–99% bacterial reads to the samples did not significantly degrade the accuracy of FunOMIC, EukDetect, or MiCoP. This suggests that marker-based methods can be reliable for datasets with bacterial background, whereas k-mer classifiers may struggle when eukaryotic reads are scarce. In real-world metatranscriptome studies, alignment-based tools have been used successfully under certain conditions. EukDetect, for example, has enabled discovery of eukaryotes in diverse environments (human gut, plant surfaces, ocean water) that were missed by standard pipelines [60]. MicroFisher, though very new, showed higher accuracy than Kraken2 or Centrifuge when tested on forest soil data [59]. Kraken2 and Centrifuge have been employed in clinical or food testing to flag potential fungal pathogens quickly, but these often require follow-up confirmation because of low specificity. In fungal microbiome surveys, tools originally developed for bacteria (Kraken2, MetaPhlAn4) are sometimes used, but their results should be interpreted cautiously.

Alignment-based classification is most effective when there is a reliable reference framework. For example, in human-associated samples where many pathogenic and commensal fungi have sequenced genomes, Kraken2 or Centrifuge can rapidly screen reads, while EukDetect can sensitively pick up low-abundance eukaryotes in metagenomes. In environmental or poorly characterized samples, marker-based profilers (EukDetect, MicroFisher, FunOMIC) are often preferred because they target conserved fungal genes. Read length also influences choice, very short reads may call for k-mer methods (Kraken2), whereas longer paired-end reads can leverage alignment to marker gene databases. In all cases, database usage is comparatively more important, with fungal-only or filtered bacterial contamination being preferred. Some pipelines use a hybrid strategy, for example, one might assemble transcripts and then run a fast classifier on contigs, or run Kraken2 on all reads to triage abundant taxa and EukDetect on unclassified reads for sensitive detection.

4. Functional classification of mycology: Lack of databases and tools

Functional classification pipelines for fungal transcriptomes typically follow one of two main strategies. Assembly-based methods first construct longer transcripts and open reading frames (ORFs) from RNA reads, which are then annotated via homology searches. This yields more complete gene models for functional assignment but requires high-quality assembly and comprehensive reference databases. In contrast, alignment-based pipelines directly map raw short reads to reference function databases (e.g., KEGG or UniProt), which avoids assembly errors but often produces only partial or ambiguous annotations because short reads may match conserved domains common to multiple proteins. Both approaches are severely constrained by sparse fungal-specific resources; many unique fungal pathways, such as secondary metabolism enzymes and CAZymes, are under-represented in general databases, and homology-based methods can even misassign fungal proteins as plant-specific factors.

4.1 Assembly-based functional classification

De novo assembly of fungal metatranscriptomes greatly enhances functional annotation by reconstructing longer transcripts that often span full-length genes or large domains [61]. Long transcripts improve ORF prediction and reveal multi-domain proteins that short reads alone would miss. In practice, RNA-Seq reads are quality-filtered and assembled (as described in Section 3.1), then processed through annotation pipelines (see **Table 3**). Common workflow steps include ORF prediction (e.g., TransDecoder to identify candidate proteins) followed by homology and domain searches. For example, Trinotate, an annotation suite, integrates BLAST/DIAMOND searches against protein databases and domain scans (via Pfam/InterProScan), then assigns gene ontology (GO) and KEGG terms to transcripts. Likewise, eggNOG-mapper [62] can be applied to assembled proteins to infer orthology-based functional categories, and InterProScan [63] identifies conserved protein families.

These approaches yield richer functional information than mapping raw reads, since full-length sequences allow complete domain profiles and correct gene models [64]. However, annotation is still limited by available databases. General-purpose resources, such as KEGG, UniProt, eggNOG, and Pfam, cover many core eukaryotic and prokaryotic genes, but lack fungal-specific pathways. For instance, fungal genomes encode unique secondary metabolism clusters (e.g., nonribosomal peptide synthases, polyketide synthases) and CAZyme repertoires for lignocellulose degradation that are under-represented in KEGG/eggNOG. Dedicated tools exist for some specialized functions, such as antiSMASH [65] for fungal secondary metabolite gene clusters and SMURF [66] for fungal secondary metabolite clusters, but these

Tool	Primary role	Function	References
TransDecoder	ORF prediction	Detects likely coding sequences in assembled transcripts using length, coding score, and domain support	[87]
Trinotate	Transcript functional annotation	Integrates BLAST/DIAMOND, Pfam/InterProScan, GO, and KEGG assignments	[88]
eggNOG-mapper	Orthology-based annotation	Assigns functions and pathways via orthologous groups; scalable to large datasets	[89]
InterProScan	Protein family and domain detection	Identifies conserved domains/motifs; integrates Pfam, SMART, TIGRFAMs	[63]
antiSMASH plus SMURF	Secondary metabolite cluster annotation	Detects fungal biosynthetic gene clusters	[65, 66]
Funannotate	Fungal genome and transcript annotation	Integrates InterProScan, eggNOG, CAZymes, antiSMASH (fungi-focused)	[90]
EnTAP	Non-model transcript annotation	Improves annotation accuracy and speed; taxonomic filtering	[88]
TIdeS	ML-based ORF prediction	Uses machine learning to improve full-length ORF prediction	[91]
TransAnnot	Modular transcript annotation	Rapid annotation using MMseqs2; integrates protein-level assembly	[92]
HPC-T-Annotator	HPC-enabled transcriptome annotation	Parallelized pipeline, web interface and Python notebook outputs	[93]
FINDER	Genome-guided transcript annotation	Integrates RNA-Seq, splicing, ORF prediction, and training with BRAKER2	[94]
CATANA	Alternative splicing annotation	Detects alternative splicing and transcription events; useful for complex fungal gene structures	[95]
AnnotaPipeline	Proteogenomics annotation	Combines RNA-Seq and mass spectrometry to validate protein-coding gene function	[96]
Annocript plus dammit	General-purpose transcript annotation	Simple end-to-end pipelines for functional assignment	[97, 98]

Table 3.
Bioinformatic tools for assembly-based functional annotation of fungal metatranscriptomic analysis.

require genome context and are not directly integrated into transcript pipelines. Similarly, mycorrhizal symbiosis pathways or fungal hormone signals are not well annotated in those databases. This database gap leads to annotations that may be non-fungal specific, for example, many fungal proteins are misannotated as plant-specific kinases or transcription factors that are absent in fungi. In other words, without fungi-focused references, homology searches can propagate errors, assigning functions (like calcium-dependent kinases or WRKY TFs) that fungi do not carry.

To tie function back to specific fungi, assembled transcripts and their predicted proteins must be taxonomically classified. The most direct strategy is homology-based, aligning each transcript or protein against databases of known fungal genomes or marker genes. Tools like DIAMOND/BLASTX can align transcripts to NCBI nr or custom fungal genome sets (e.g., JGI MycoCosm), and then assign taxonomy via LCA algorithms. For instance, Peng and Valentine [67] aligned metatranscriptome genes to nr with DIAMOND and used BASTA-LCA to classify sequences to fungal phyla. Alternatively, marker-gene approaches scan for conserved single-copy genes within transcripts and match these to fungal clades. In any case, because environmental samples contain mixed microbes, many transcripts produce ambiguous hits. High-confidence "bins" or clusters help mitigate this. First, cluster contigs by nucleotide composition or coverage to group fragments from the same organism, and then annotating the entire bin would be a strategy to classify the *de novo* assembled transcripts. Combining rRNA marker assignment with assembled mRNA transcript annotation will improve the overall classification confidence. Schneider et al. [8] filtered out host plant sequences, assembled fungal transcripts, and then jointly annotated taxonomy and function. They showed that despite sparse reference genomes, combining ITS amplicon data with transcriptome assembly improved confidence in fungal assignments. It is a common practice to assign a contig to single fungi with low confidence, but if all transcripts in a bin share the same best-hit species, it is a more reliable attribute that groups functions belong to a specific fungus.

Linking taxonomy to function can reveal fungal ecological roles. Auer et al. [68] classified transcripts into broad guilds (saprotrophs, pathogens, ectomycorrhizal) using reference genomes, and then compared their expression profiles. They found that saprotrophic fungi in soil up-regulate plant cell wall-degrading enzymes relative to mycorrhizal fungi, whereas ectomycorrhizal fungi up-regulate nitrogen transporters. This kind of analysis depends on correctly binning transcripts to taxa, but it can discern, for example, whether lignin-degrading laccases come from wood-decayers (saprotrophs) or from symbionts. In another study, contigs annotated as *Rhizoctonia*, a plant pathogen, encoded ligninases in the assembled metatranscriptome, directly linking that fungus to lignin degradation [69]. In summary, strategies that assign assembled transcripts to fungal taxa enable an integrated view of "who is doing what" in a community. Challenges remain, but hybrid approaches that iterate binning, annotation, and expression profiling are increasingly used to trace functions back to specific fungi. *De novo* assembly-based annotation allows coupling gene expression with pathway analysis. Assembled transcripts can be quantified across samples to infer active pathways under different conditions. For instance, Hayden et al. [69] compared wheat rhizosphere soils and showed that assembly recovered many differentially expressed metabolic genes missed by raw-read methods. In suppressive soils, assembled transcripts included a polyketide cyclase and a terpenoid backbone enzyme, whereas non-suppressive soils showed overexpression of antibiotic biosynthesis genes (e.g., phenazine pathway). Similarly, in plant–fungal symbioses, assembly has revealed key fungal activities. These case studies show that assembly-based annotation can link fungal gene expression to ecological function by providing complete gene sequences and enabling quantification of pathway activities.

4.2 Alignment-based functional classification: Challenges and future directions

An alternative to assembly is direct functional annotation of raw reads through alignment-based approach. Tools like HUMAnN3 [70], MEGAN [71], or FMAP [72] accept unassembled reads and align them to functional databases (KEGG, SEED, UniRef) to profile community functions. While these methods avoid potential assembly errors, they face serious limitations with fungal data. Short reads (150 bp) often hit only one domain of a multi-domain fungal enzyme, yielding partial or spurious annotations. For example, MEGAN can count how many reads hit a KEGG ortholog, but a read matching a conserved enzyme fragment cannot confirm the presence of full genes. Moreover, because many microbial functions, such as cellulase activity or glycolysis, are shared by bacteria and fungi, read-based functional annotations do not distinguish source, this means a glycosyl hydrolase read could be from a fungus or a bacterium, confounding ecological interpretation. Hayden et al. [69] applied assembly- and alignment-based approaches to wheat rhizosphere metatranscriptomic and found that the assembled pipeline identified many more differentially expressed genes than the unassembled MEGAN-based pipeline. In their study, unassembled analysis relied on MEGAN assigning reads to COG/KEGG terms, but still missed dozens of gene families that the assembly approach captured. Notably, most highly expressed contigs in that soil dataset turned out to be bacterial (over 80% of contigs), meaning fungal transcripts were underrepresented without assembly. Similarly, Stewart et al. [73] report that read alignment-based methods often recover only the most abundant taxa and miss low-expression eukaryotic genes. In short, functional classification from unassembled reads tends to be biased toward well-represented bacterial genes, and yields many "unassigned" or "hypothetical" hits for fungal transcripts.

To address those issues, several works could improve alignment-based fungal annotation. First, fungal-enriched reference databases are needed. Expanding databases of fungal protein families, including genes and pathways from fungi, would allow mapping reads to fungal metabolic pathways. For instance, CAZy database catalogs fungal CAZymes, and efforts to integrate CAZy families into metatranscriptomic tools would help recognize fungal polysaccharide-degrading functions. Similarly, specialized resources like antiSMASH (fungal) and SMURF highlight secondary metabolite genes (polyketides, nonribosomal peptides) that are missed in general databases. Incorporating outputs of these tools into metatranscriptomic analysis could reveal fungal activity. Second, machine-learning approaches may assign functions without direct homologs. Fungal genomes have unique protein families, and machine-learning models trained on fungal sequences can help predict novel functions. Deep-learning models could be adapted with fungal training sets. Efforts to build fungal-specific HMMs or neural nets for gene families (CAZymes, effector proteins) would complement homology searches. Finally, community curation and standards are essential. Fungi deserve dedicated annotation efforts analogous to those for bacteria. The DOE JGI's MycoCosm project, as a leading example, employs an automated pipeline plus manual curation to annotate fungal genomes. Standard operating procedures ensure that fungal gene models and functions are consistently predicted and reviewed. Likewise, community resources such as FungiDB and GO's fungal annotations help consolidate knowledge of fungal pathways. Collaborative initiatives to curate fungal-specific gene families (CAZymes, effectors, symbiosis genes) and to integrate them into public databases will improve all downstream metatranscriptome analyses.

5. Concluding remarks

Metatranscriptomics is revolutionizing mycology by simultaneously revealing which fungi are present and actively expressing genes, and what metabolic processes they are performing. RNA-Seq of environmental samples targets the expressed transcript pool, offering a "live" snapshot of the molecular function. By linking expressed genes to specific fungi, metatranscriptomics enables a true "who does what" view of the mycobiome, which has begun to tease apart fungal lifestyles. These integrated taxonomy–function profiles can inform ecosystem models, since microbial-mediated decomposition and nutrient cycling depend on both who is there and what genes they express. Fungal transcriptomes are inherently complex, many fungal genes contain introns and generate alternative isoforms, which complicates assembly and quantification. Common metagenome assemblers may misassemble spliced fungal transcripts. Environmental RNA samples are also dominated by non-fungal material (host, plant, bacteria), so bioinformatic filtering is needed to recover the fungal signal. In addition, a lack of fungal references results in the limitation to comprehensively interpret metatranscriptomic datasets. Most available genomes are biased toward model Ascomycota and Basidiomycota, so many environmental fungi, especially basal lineages and cryptic symbionts, lack close matches. Thus, mycobiome profiling tools are still few, and none fully cover fungal diversity. On the functional side, widely used databases (KEGG, eggNOG, UniProt) are built mainly on bacteria and model eukaryotes, so novel fungal enzymes, such as unique CAZymes, secondary metabolite synthases, and effectors, often have no annotation. Consequently, many transcripts are labeled "hypothetical protein" or mislabeled with distant plant/bacterial homologs. In short, current tools capture mostly housekeeping and conserved pathways, missing the unique aspects of fungal metabolism.

In conclusion, advanced sequencing and analysis are finally unveiling the hidden dynamics of fungal communities. Metatranscriptomics bridges the gap between composition and function, it shows not only "what fungi are out there" but "what they are doing" right now. Although challenges of assembly, contamination, and annotation persist, the field is rapidly evolving. These tools are transforming our understanding of the mycobiome. The metatranscriptomic makes it possible to track how fungal communities respond to environmental change, interact with hosts, and drive biogeochemical cycles. With continued improvements in long-read data, comprehensive reference genomes, and community standards, fungal metatranscriptomics will provide ever more accurate and reproducible portraits of the active fungal world.

Funding

This research was funded by Natural Science Foundation of China (grant no. 32400097).

Author details

Haihua Wang
Department of Soil, Water, and Ecosystem Sciences, University of Florida, Gainesville, FL, USA

*Address all correspondence to: wanghaihua@ufl.edu

References

[1] Hawksworth DL, Lücking R. Fungal diversity revisited: 2.2 to 3.8 million species. Microbiology Spectrum. 2017;5. DOI: 10.1128/microbiolspec.FUNK-0052-2016

[2] Debeljak P, Baltar F. Fungal diversity and community composition across ecosystems. Journal of Fungi (Basel, Switzerland). 2023;**9**. DOI: 10.3390/jof9050510

[3] Watkinson SC. Mutualistic symbiosis between fungi and autotrophs. In: The Fungi. Elsevier; 2016. p. 205–243

[4] Nilsson RH, Anslan S, Bahram M, Wurzbacher C, Baldrian P, Tedersoo L. Mycobiome diversity: High-throughput sequencing and identification of fungi. Nature Reviews Microbiology. 2019;**17**:95–109

[5] Qin Y, Wu L, Zhang Q, Wen C, Van Nostrand JD, Ning D, et al. Effects of error, chimera, bias, and GC content on the accuracy of amplicon sequencing. mSystems. 2023;**8**:e0102523

[6] Johnson SM, Carlson EL, Pappagianis D. Determination of ribosomal DNA copy number and comparison among strains of Coccidioides. Mycopathologia. 2015;**179**:45–51

[7] Cottier F, Srinivasan KG, Yurieva M, Liao W, Poidinger M, Zolezzi F, et al. Advantages of meta-total RNA sequencing (MeTRS) over shotgun metagenomics and amplicon-based sequencing in the profiling of complex microbial communities. Npj Biofilms and Microbiomes. 2018;**4**:2

[8] Schneider AN, Sundh J, Sundström G, Richau K, Delhomme N, Grabherr M, et al. Comparative fungal community analyses using metatranscriptomics and internal transcribed spacer amplicon sequencing from Norway spruce. mSystems. 2021:6. DOI: 10.1128/mSystems.00884-20

[9] Aplakidou E, Vergoulidis N, Chasapi M, Venetsianou NK, Kokoli M, Panagiotopoulou E, et al. Visualizing metagenomic and metatranscriptomic data: A comprehensive review. Computational and Structural Biotechnology Journal. 2024;**23**:2011–2033

[10] Mao R, Li X, Ge M, Cambria E. MetaPro: A computational metaphor processing model for text pre-processing. Information Fusion. 2022;**86-87**:30–43

[11] Martinez X, Pozuelo M, Pascal V, Campos D, Gut I, Gut M, et al. MetaTrans: An open-source pipeline for metatranscriptomics. Scientific Reports. 2016:6. DOI: 10.1038/srep26447

[12] Xie Z, Manichanh C. FunOMIC: Pipeline with built-in fungal taxonomic and functional databases for human mycobiome profiling. Computational and Structural Biotechnology Journal. 2022;**20**:3685–3694

[13] Hesse CN, Mueller RC, Vuyisich M, Gallegos-Graves LV, Gleasner CD, Zak DR, et al. Forest floor community metatranscriptomes identify fungal and bacterial responses to N deposition in two maple forests. Frontiers in Microbiology. 2015;**6**:337

[14] Zeng Q, Lebreton A, Auer L, Man X, Jia L, Wang G, et al. Stable functional structure despite high taxonomic variability across fungal communities in soils of old-growth montane forests. Microbiome. 2023;**11**:217

[15] Marcelino VR, Irinyi L, Eden J-S, Meyer W, Holmes EC, Sorrell TC. Metatranscriptomics as a tool to identify fungal species and subspecies in mixed communities – A proof of concept under laboratory conditions. IMA Fungus. 2019;**10**:12

[16] Carvalhais LC, Dennis PG, Tyson GW, Schenk PM. Application of metatranscriptomics to soil environments. Journal of Microbiological Methods. 2012;**91**: 246–251

[17] Sharuddin SS, Ramli N, Yusoff MZM, Muhammad NAN, Ho LS, Maeda T. Advancement of metatranscriptomics towards productive agriculture and sustainable environment: A review. International Journal of Molecular Sciences. 2022;**23**:3737

[18] Tavares S, Ramos AP, Pires AS, Azinheira HG, Caldeirinha P, Link T, et al. Genome size analyses of Pucciniales reveal the largest fungal genomes. Frontiers in Plant Science. 2014;**5**:422

[19] Orozco-Arias S, Candamil MS, Jaimes PA, Cristancho M, Tabares-Soto R, Guyot R. Composition and diversity of LTR retrotransposons in the coffee leaf rust genome (Hemileia vastatrix). Agronomy (Basel). 2022;**12**:1665

[20] Cristancho MA, Botero-Rozo DO, Giraldo W, Tabima J, Riaño-Pachón DM, Escobar C, et al. Annotation of a hybrid partial genome of the coffee rust (Hemileia vastatrix) contributes to the gene repertoire catalog of the Pucciniales. Frontiers in Plant Science. 2014;**5**:594

[21] Jeon J, Kim K-T, Choi J, Cheong K, Ko J, Choi G, et al. Alternative splicing diversifies the transcriptome and proteome of the rice blast fungus during host infection. RNA Biology. 2022;**19**:373–385

[22] Zhong Z, Chen M, Lin L, Chen R, Liu D, Norvienyeku J, et al. Genetic variation bias toward noncoding regions and secreted proteins in the rice blast fungus Magnaporthe oryzae. mSystems. 2020:5. DOI: 10.1128/mSystems. 00346-20

[23] Brauer EK, Subramaniam R, Harris LJ. Regulation and dynamics of gene expression during the life cycle of Fusarium graminearum. Phytopathology. 2020;**110**:1368–1374

[24] Tan K, Wong KH. RNA polymerase II ChIP-seq—a powerful and highly affordable method for studying fungal genomics and physiology. Biophysical Reviews. 2019;**11**:79–82

[25] Fradin C, Kretschmar M, Nichterlein T, Gaillardin C, d'Enfert C, Hube B. Stage-specific gene expression of Candida albicans in human blood. Molecular Microbiology. 2003;**47**:1523–1543

[26] Avershina E, Qureshi AI, Winther-Larsen HC, Rounge TB. Challenges in capturing the mycobiome from shotgun metagenome data: Lack of software and databases. Microbiome. 2025;**13**:66

[27] Gryganskyi AP, Golan J, Muszewska A, Idnurm A, Dolatabadi S, Mondo SJ, et al. Sequencing the

genomes of the first terrestrial fungal lineages: What have we learned? Microorganisms. 2023:11. DOI: 10.3390/microorganisms11071830

[28] Muggia L, Ametrano CG, Sterflinger K, Tesei D. An overview of genomics, phylogenomics and proteomics approaches in ascomycota. Life (Basel, Switzerland). 2020;**10**:356

[29] Kanehisa M, Furumichi M, Tanabe M, Sato Y, Morishima K. KEGG: New perspectives on genomes, pathways, diseases and drugs. Nucleic Acids Research. 2017;**45**:D353–D361

[30] Tatusov RL. The COG database: A tool for genome-scale analysis of protein functions and evolution. Nucleic Acids Research. 2000;**28**:33–36

[31] Mistry J, Chuguransky S, Williams L, Qureshi M, Salazar GA, Sonnhammer ELL, et al. Pfam: The protein families database in 2021. Nucleic Acids Research. 2021;**49**: D412–D419

[32] Song Y, Wan J, Shang -J-J, Feng Z, Jin Y, Li H, et al. The complete mitochondrial genome of the edible mushroom Grifola frondosa. Mitochondrial DNA Part B. 2022;7: 286–288

[33] Li D, C-m L, Luo R, Sadakane K, Lam T-W. MEGAHIT: An ultra-fast single-node solution for large and complex metagenomics assembly via succinct de Bruijn graph. Bioinformatics. 2015;**31**:1674–1676

[34] Nurk S, Meleshko D, Korobeynikov A, Pevzner PA. metaSPAdes: A new versatile metagenomic assembler. Genome Research. 2017;**27**:824–834

[35] LaPierre N, Mangul S, Alser M, Mandric I, Wu NC, Koslicki D, et al. MiCoP: Microbial community profiling method for detecting viral and fungal organisms in metagenomic samples. BMC Genomics. 2019;**20**:423

[36] Liu Y, Ghaffari MH, Ma T, Tu Y. Impact of database choice and confidence score on the performance of taxonomic classification using Kraken2. Abiotech. 2024;**5**:465–475

[37] Kim D, Song L, Breitwieser FP, Salzberg SL. Centrifuge: Rapid and sensitive classification of metagenomic sequences. Genome Research. 2016;**26**:1721–1729

[38] Blanco-Míguez A, Beghini F, Cumbo F, McIver LJ, Thompson KN, Zolfo M, et al. Extending and improving metagenomic taxonomic profiling with uncharacterized species using MetaPhlAn 4. Nature Biotechnology. 2023;**41**:1633–1644

[39] Tedersoo L, Albertsen M, Anslan S, Callahan B. Perspectives and benefits of high-throughput long-read sequencing in microbial ecology. Applied and Environmental Microbiology. 2021;**87**: e0062621

[40] Oikonomopoulos S, Bayega A, Fahiminiya S, Djambazian H, Berube P, Ragoussis J. Methodologies for transcript profiling using long-read technologies. Frontiers in Genetics. 2020;**11**:606

[41] Zhao L, Zhang H, Kohnen MV, Kvsk P, Gu L, Reddy ASN. Analysis of transcriptome and epitranscriptome in plants using PacBio Iso-Seq and nanopore-based direct RNA sequencing. Frontiers in Genetics. 2019;**10**:253

[42] Grigoriev IV, Nikitin R, Haridas S, Kuo A, Ohm R, Otillar R, et al. MycoCosm portal: Gearing up for 1000 fungal genomes. Nucleic Acids Research. 2014;**42**:D699–704

[43] Basenko EY, Pulman JA, Shanmugasundram A, Harb OS, Crouch K, Starns D, et al. FungiDB: An integrated bioinformatic resource for fungi and Oomycetes. Journal of Fungi (Basel, Switzerland). 2018:4. DOI: 10.3390/jof4010039

[44] Singh A, Vats S, Bhargava P. Advances and Challenges in Metatranscriptomic Analysis. Microbial Metatranscriptomics Belowground. Singapore: Springer Singapore; 2021. p. 453–469

[45] Shakya M, Lo -C-C, Chain PSG. Advances and challenges in metatranscriptomic analysis. Frontiers in Genetics. 2019;**10**:904

[46] Lind D, Marcus B. An Introduction to Symbolic Dynamics and Coding an Introduction to Symbolic Dynamics and Coding. 2nd ed. Cambridge, England: Cambridge University Press (Virtual Publishing); 2021

[47] Haas BJ, Papanicolaou A, Yassour M, Grabherr M, Blood PD, Bowden J, et al. De novo transcript sequence reconstruction from RNA-seq using the Trinity platform for reference generation and analysis. Nature Protocols. 2013;**8**:1494–1512

[48] Schulz MH, Zerbino DR, Vingron M, Birney E. Oases: Robust de novo RNA-seq assembly across the dynamic range of expression levels. Bioinformatics. 2012;**28**:1086–1092

[49] Bushmanova E, Antipov D, Lapidus A, Prjibelski AD. rnaSPAdes: A de novo transcriptome assembler and its application to RNA-Seq data. Gigascience. 2019;8. DOI: 10.1093/gigascience/giz100

[50] Robertson G, Schein J, Chiu R, Corbett R, Field M, Jackman SD, et al. De novo assembly and analysis of RNA-seq data. Nature Methods. 2010;7:909–912

[51] Xie Y, Wu G, Tang J, Luo R, Patterson J, Liu S, et al. SOAPdenovo-Trans: De novo transcriptome assembly with short RNA-Seq reads. Bioinformatics. 2014;**30**:1660–1666

[52] Liu J, Li G, Chang Z, Yu T, Liu B, McMullen R, et al. BinPacker: Packing-based DE Novo transcriptome assembly from RNA-seq data. PLOS Computational Biology. 2016;**12**: e1004772

[53] Wang S, Gribskov M. Comprehensive evaluation of de novo transcriptome assembly programs and their effects on differential gene expression analysis. Bioinformatics. 2017;**33**:327–333

[54] Bucchini F, Del Cortona A, Kreft Ł, Botzki A, Van Bel M, Vandepoele K. TRAPID 2.0: A web application for taxonomic and functional analysis of de novo transcriptomes. Nucleic Acids Research. 2021;**49**:e101.

[55] Freedman AH, Clamp M, Sackton TB. Error, noise and bias in de novo transcriptome assemblies. Molecular Ecology Resources. 2021;**21**:18–29

[56] Celaj A, Markle J, Danska J, Parkinson J. Comparison of assembly algorithms for improving rate of metatranscriptomic functional annotation. Microbiome. 2014;**2**:39

[57] Wood DE, Lu J, Langmead B. Improved metagenomic analysis with Kraken 2. Genome Biology. 2019;**20**:257

[58] Menzel P, Ng KL, Krogh A. Fast and sensitive taxonomic classification for metagenomics with Kaiju. Nature Communications. 2016;7(1):1257

[59] Wang H, Wu S, Zhang K, Chen K-H, Vilgalys R, Liao H-L. MicroFisher: Fungal taxonomic classification for metatranscriptomic and metagenomic data using multiple short hypervariable markers. bioRxiv. 2024;2024.01.20.576350. DOI: 10.1101/2024.01.20.576350

[60] Lind AL, Pollard KS. Accurate and sensitive detection of microbial eukaryotes from whole metagenome shotgun sequencing. Microbiome. 2021;**9**:58

[61] Raghavan V, Kraft L, Mesny F, Rigerte L. A simple guide to de novo transcriptome assembly and annotation. Briefings in Bioinformatics. 2022;23. DOI: 10.1093/bib/bbab563

[62] Hernández-Plaza A, Szklarczyk D, Botas J, Cantalapiedra CP, Giner-Lamia J, Mende DR, et al. eggNOG 6.0: Enabling comparative genomics across 12 535 organisms. Nucleic Acids Research. 2023;**51**:D389–D394

[63] Jones P, Binns D, Chang H-Y, Fraser M, Li W, McAnulla C, et al. InterProScan 5: Genome-scale protein function classification. Bioinformatics. 2014;**30**:1236–1240

[64] Miryeganeh M, Saze H. De Novo transcriptome assembly, functional annotation, and transcriptome dynamics analyses reveal stress tolerance genes in mangrove tree (Bruguiera gymnorhiza). International Journal of Molecular Sciences. 2021;**22**:9874

[65] Blin K, Shaw S, Kloosterman AM, Charlop-Powers Z, van Wezel GP, Medema MH, et al. antiSMASH 6.0: Improving cluster detection and comparison capabilities. Nucleic Acids Research. 2021;**49**:W29–W35

[66] Khaldi N, Seifuddin FT, Turner G, Haft D, Nierman WC, Wolfe KH, et al. SMURF: Genomic mapping of fungal secondary metabolite clusters. Fungal Genetics and Biology. 2010;**47**:736–741

[67] Peng X, Valentine DL. Expanded genetic and functional diversity of oceanic fungi. Microbiome. 2025;**13**:179

[68] Auer L, Buée M, Fauchery L, Lombard V, Barry KW, Clum A, et al. Metatranscriptomics sheds light on the links between the functional traits of fungal guilds and ecological processes in forest soil ecosystems. New Phytologist. 2024;**242**:1676–1690

[69] Hayden HL, Savin KW, Wadeson J, Vvsr G, Mele PM. Comparative metatranscriptomics of wheat rhizosphere microbiomes in disease suppressive and non-suppressive soils for Rhizoctonia solani AG8. Frontiers in Microbiology. 2018;**9**:859

[70] Franzosa EA, McIver LJ, Rahnavard G, Thompson LR, Schirmer M, Weingart G, et al. Species-level functional profiling of metagenomes and metatranscriptomes. Nature Methods. 2018;**15**:962–968

[71] Bağcı C, Beier S, Górska A, Huson DH. Introduction to the analysis of environmental sequences: Metagenomics with MEGAN. Methods in Molecular Biology. 2019;**1910**: 591–604

[72] Kim J, Kim MS, Koh AY, Xie Y, Zhan X. FMAP: Functional Mapping and Analysis Pipeline for metagenomics and metatranscriptomics studies. BMC Bioinformatics. 2016;**17**:420

[73] Stewart FJ, Sharma AK, Bryant JA, Eppley JM, DeLong EF. Community transcriptomics reveals universal patterns of protein sequence conservation in natural microbial communities. Genome Biology. 2011;**12**:R26

[74] Peay KG, Kennedy PG, Talbot JM. Dimensions of biodiversity in the Earth mycobiome. Nature Reviews Microbiology. 2016;**14**:434–447

[75] Grabherr MG, Haas BJ, Yassour M, Levin JZ, Thompson DA, Amit I, et al. Full-length transcriptome assembly from RNA-Seq data without a reference genome. Nature Biotechnology. 2011;**29**:644–652

[76] Buchfink B, Xie C, Huson DH. Fast and sensitive protein alignment using DIAMOND. Nature Methods. 2015;**12**:59–60

[77] Pearson WR. BLAST and FASTA similarity searching for multiple sequence alignment. Methods in Molecular Biology. 2014;**1079**:75–101

[78] Kahlke T, Ralph PJ. BASTA – Taxonomic classification of sequences and sequence bins using last common ancestor estimations. Methods in Ecology and Evolution. 2019;**10**: 100–103

[79] Soverini M, Turroni S, Biagi E, Brigidi P, Candela M, Rampelli S. HumanMycobiomeScan: A new bioinformatics tool for the characterization of the fungal fraction in metagenomic samples. BMC Genomics. 2019;**20**:496

[80] Donovan PD, Gonzalez G, Higgins DG, Butler G, Ito K. Identification of fungi in shotgun metagenomics datasets. PLoS One. 2018;**13**:e0192898

[81] Huson DH, Auch AF, Qi J, Schuster SC. MEGAN analysis of metagenomic data. Genome Research. 2007;**17**:377–386

[82] Marcelino VR, Clausen PTLC, Buchmann JP, Wille M, Iredell JR, Meyer W, et al. CCMetagen: Comprehensive and accurate identification of eukaryotes and prokaryotes in metagenomic data. Genome Biology. 2020;**21**:103

[83] Dröge J, Gregor I, McHardy AC. Taxator-tk : Precise taxonomic assignment of metagenomes by fast approximation of evolutionary neighborhoods. Bioinformatics. 2015;**31**:817–824

[84] Espinoza JL, Dupont CL. VEBA: A modular end-to-end suite for in silico recovery, clustering, and analysis of prokaryotic, microeukaryotic, and viral genomes from metagenomes. BMC Bioinformatics. 2022;**23**:419

[85] Milanese A, Mende DR, Paoli L, Salazar G, Ruscheweyh H-J, Cuenca M, et al. Microbial abundance, activity and population genomic profiling with mOTUs2. Nature Communications. 2019;**10**:1014

[86] Krinos A, Hu S, Cohen N, Alexander H. EUKulele: Taxonomic annotation of the unsung eukaryotic microbes. Journal of Open Source Software. 2021;**6**:2817

[87] Mao A, Ji HJ, Haas B, Salzberg S, Sommer MJ. TD2: Finding protein coding regions in transcripts. bioRxiv : the Preprint Server for Biology. 2025. DOI: 10.1101/2025.04.13.648579

[88] Hart AJ, Ginzburg S, Xu MS, Fisher CR, Rahmatpour N, Mitton JB, et al. EnTAP: Bringing faster and smarter functional annotation to non-model eukaryotic transcriptomes. Molecular Ecology Resources. 2020;**20**:591–604

[89] Cantalapiedra CP, Hernández-Plaza A, Letunic I, Bork P, Huerta-Cepas J. eggNOG-mapper v2: Functional Annotation, Orthology Assignments, and Domain Prediction at the Metagenomic Scale. Molecular Biology and Evolution. 2021;**38**:5825–5829.

[90] Palmer JM, Stajich J. Funannotate v1.8.1: Eukaryotic genome annotation. Zenodo. 2020. DOI: 10.5281/ZENODO.1134477

[91] Xx M-A, Kim E. TIdeS: A comprehensive framework for accurate open reading frame identification and classification in eukaryotic transcriptomes. Genome Biology and Evolution. 2024;16. DOI: 1 0.1093/gbe/evae252

[92] Zelenskaia M, Arangasamy Y, Mirdita M, Söding J, Raghavan V. TransAnnot-a fast transcriptome annotation pipeline. Bioinformatics Advances. 2024;**4**:vbae152

[93] Arcioni L, Arcieri M, Martino JD, Liberati F, Bottoni P, Castrignanò T. HPC-T-Annotator: An HPC tool for de novo transcriptome assembly annotation. BMC Bioinformatics. 2024;**25**:272

[94] Banerjee S, Bhandary P, Woodhouse M, Sen TZ, Wise RP, Andorf CM. FINDER: An automated software package to annotate eukaryotic genes from RNA-Seq data and associated protein sequences. BMC Bioinformatics. 2021;**22**:205

[95] Shiau C-K, Huang J-H, Tsai H-K. CATANA: A tool for generating comprehensive annotations of alternative transcript events. Bioinformatics. 2019;**35**:1414–1415

[96] Maia GA, Filho VB, Kawagoe EK, Teixeira Soratto TA, Moreira RS, Grisard EC, et al. AnnotaPipeline: An integrated tool to annotate eukaryotic proteins using multi-omics data. Frontiers in Genetics. 2022;**13**:1020100. DOI: 10.3389/fgene.2022.1020100

[97] Musacchia F, Basu S, Petrosino G, Salvemini M, Sanges R. Annocript: A flexible pipeline for the annotation of transcriptomes able to identify putative long noncoding RNAs. Bioinformatics. 2015;**31**:2199–2201

[98] Scott C, Pierce T, Johnson LK, Titus Brown C, Reid C. The Gitter Badger, et al. dib-lab/dammit: Dammit v1.2. Zenodo. 2019. DOI: 10.5281/ZENODO.3569831

Chapter 5

Novel Biotechnological Methods for the Cultivation and Inoculation of Arbuscular Mycorrhizal Fungi

Irfan Ali, Taqwa Saeed, Rimsha Riaz, Rao Sohail Ahmed Khan, Ghulam Mustafa, Ahsan Iqbal, Tatheer Alam Naqvi and Adnan Khan Niazi

Abstract

Arbuscular mycorrhizal fungi (AMF) are transitioning toward scalable biostimulants despite various challenges. This chapter redefines AMF production using soil-less, sterile pipelines that initiate with soil-free root organ culture (ROC: transformed roots used for soil-less contaminant-controlled propagation) and advance to low-shear, O_2-enriched bioreactors designed for process regulation and harvesting. In this study, recent advancements in encapsulation and stabilization, specifically alginate beads, film coating, and cryopreservation, are discussed to facilitate dosage precision, moisture regulation, extended shelf life, and traceability *via* master bank systems. A realistic standardized quality framework is proposed that assesses identity and purity, viability, and function through infective propagule (IP) bioassays. The soil free culture was synthesized on the transformed roots in the customized bioreactor that enabled the propagation of propagules under the aseptic conditions. This helps in improving productivity and standardizing quality. We also emphasized the cryopreservation and encapsulation strategies to explain the role of alginate and other biodegradable carriers in maintaining viability. It can ensure the storage and stability of strains for a longer duration.

We discussed the principles of engineering of formulations along with molecular methods. We linked these components and design principles for AMF and mycorrhiza-helper-bacterial consortia. However, in the techno-economic analysis, some cost determinants were identified. We emphasized uniform measurements to ensure that the performance assertions are reproducible across various crops and locations. It is also discussed to enhance the role of beneficial microbes to be used as co-inoculants. It can increase their efficiency in field particularlry in the abiotic conditions. The result of this chapter suggests bridging the gap in the lbal research and economical field formulations.

Keywords: AMF, encapsulation, ROC, quality control, cryopreservation, bioreactors

1. Introduction

The transition of AMF inoculants has taken place from specialized tools to widely used biostimulants. However, inconsistency in the product remains an important barrier to its adoption. There are biological and industrial factors that play important roles in standardization, variability, and quality. Contemporary ROC systems use *Agrobacterium rhizogenes,* which yields genetically stable and axenic host tissues that facilitate AMF proliferation in the absence of soil, thereby allowing for process regulation and contaminant prevention (**Figure 1**). Recent efforts have focused on establishing standardized criteria for optimizing yield in ROC and laying the groundwork for industrialization [1–4].

Simultaneously, regulators establish the minimum quality and safety standards. The EU Fertilizing Goods Regulation (EU) 2019/1009 establishes labeling and microbiological standards for microbial goods, whereas many nations (e.g., India) delineate minimum spore or infective-propagule thresholds for AMF products. These frameworks compel manufacturers to achieve quantifiable, replicable quality and substantiated assertions [5–7].

In this context, the current chapter advocates for a manufacturing-oriented re-evaluation of AMF culture and inoculation: transitioning from diverse, soil-dependent "trap culture" to soil-free, quality-controlled, and commercially viable procedures. We deliberately integrated biology and engineering. From a biological perspective, we evaluated species or strain selection, host root lines, and construction of consortia, including mycorrhiza-helper bacteria (MHB). We established an engineering pipeline that initiates with ROC in controlled containers and concludes with unit operations for harvesting to delivery, with each phase linked to release requirements and shelf-life objectives [1, 7].

Conventional substrate-based AMF manufacturing possesses three structural deficiencies that contemporary biotechnology can address. Contamination control is inherently challenging when soil or compost serves as the growth medium, resulting in inconsistent purity and the presence of unintended bacteria that complicate efficacy and safety evaluations. Second, the measurement of processes is constrained; spore counts alone fail to reflect infectivity or vitality, and variations in root colonization across batches sometimes remain undetected until field trials are conducted. Third, low propagule density and bulky carriers hinder the scale and logistics, leading to elevated transport and application expenses [3, 8, 9]. ROC-based technologies directly mitigate the initial two deficiencies by offering sterile, traceable cultures; integrating ROC with bioreactors and encapsulation remedies the third deficiency by enhancing volumetric productivity and facilitating high-potency, manageable forms [8].

The primary biological limitation is that AMF are obligate symbionts that require living roots for growth. ROC fulfills this criterion while enabling the engineering of media composition. The risk of contamination and low-cost scaling up is addressed by the integration of ROC with low-shear, high-oxygen bioreactors [10, 11].

Formulation science is also an important pillar of dependability. AMF propagules exhibit sensitivity to thermal conditions and desiccation, resulting in diminished vigor when subjected to elevated oxygen levels during storage. Encapsulation within biodegradable matrices safeguards propagules during desiccation and manipulation, regulates moisture content, and facilitates regulated release in the rhizosphere. The use of benign protectants, such as trehalose or proteinaceous excipients, along with barrier packaging, including moisture and oxygen scavengers, can prolong shelf life while maintaining infectivity. Encapsulation-assisted cryopreservation facilitates a master-cell-bank strategy, similar to industrial microbe production. This renders batch genealogy verifiable and reduces trait drift from successive subculturing [12–16].

The reframing of quality control is necessary. First, we check for identity and purity (using morphology and molecular markers to confirm that there are no specific contaminants), then we check for viability, then we check for potency (using a standardized most-probable-number bioassay to measure the number of infectious propagules per gram), and finally, we check for stability (ensuring that the virus remains active under certain storage conditions). These values can be built as controls during the process (for example, to check for infectivity during breaks before drying) or as release tests linked to claims on the label. Aligning internal standards with new regulatory standards ensures that both customers and certifiers can understand the effectiveness of data created in trials [17–21].

The dose, placement, time, and compatibility determine how well something works in the field. We focused on delivery methods because AMF colonization is most helpful when it starts early. These include seed coatings for small-seed crops and veggie transplants, in-furrow gels or granules for broadacre crops, and plug/root dips for nurseries (**Figure 2**). Growers should also be given practical advice on how to water and avoid salt and fungicide exposure in the crucial first two to three weeks after application. When it makes sense, mixing or delivering MHB (like certain types of Bacillus and Pseudomonas) with the main ingredient can help the plant grow and take in nutrients better, as long as the pH and water activity of the carrier are good for both partners [22–25].

1.1 Structure and scope of chapter

This opening explains why AMF inoculants should be examined from the point of view of the process. That is all in Section 2, which goes over soil-free growth, from the basics of ROC to choosing the right reactor and how to scale it up. We discuss the approaches from encapsulation to cryopreservation to make food last longer. In the next sections, formulation engineering is made official, and a quality control toolkit is put together that meets legal requirements. We also discuss inoculation technologies and provide realistic suggestions for using them in the field. We discuss the AMF–bacterial consortia and create rules for making them work together. The next section combines economics and manufacturing, turning lab protocols into strong output. Finally, we propose a future plan that will make it easier to compare things (for example, the number of infective propagules per unit cost) and promote clear labeling and cost-effective goods for growers [4, 9].

2. AMF culture: from ROC to bioreactors

AMF cannot be grown from pure cultures; they require roots that are still alive. Root-organ culture (ROC) addresses this issue by introducing AMF and *Agrobacterium rhizogenes* onto specific media. This method creates axenic and repeatable systems for propagating plants and studying their functions. Early research revealed that the fungus infected Ri roots, but it was unclear how the spores were spreading. A significant breakthrough occurred when carrot ROC produced a large number of viable spores, providing a simpler solution. This development

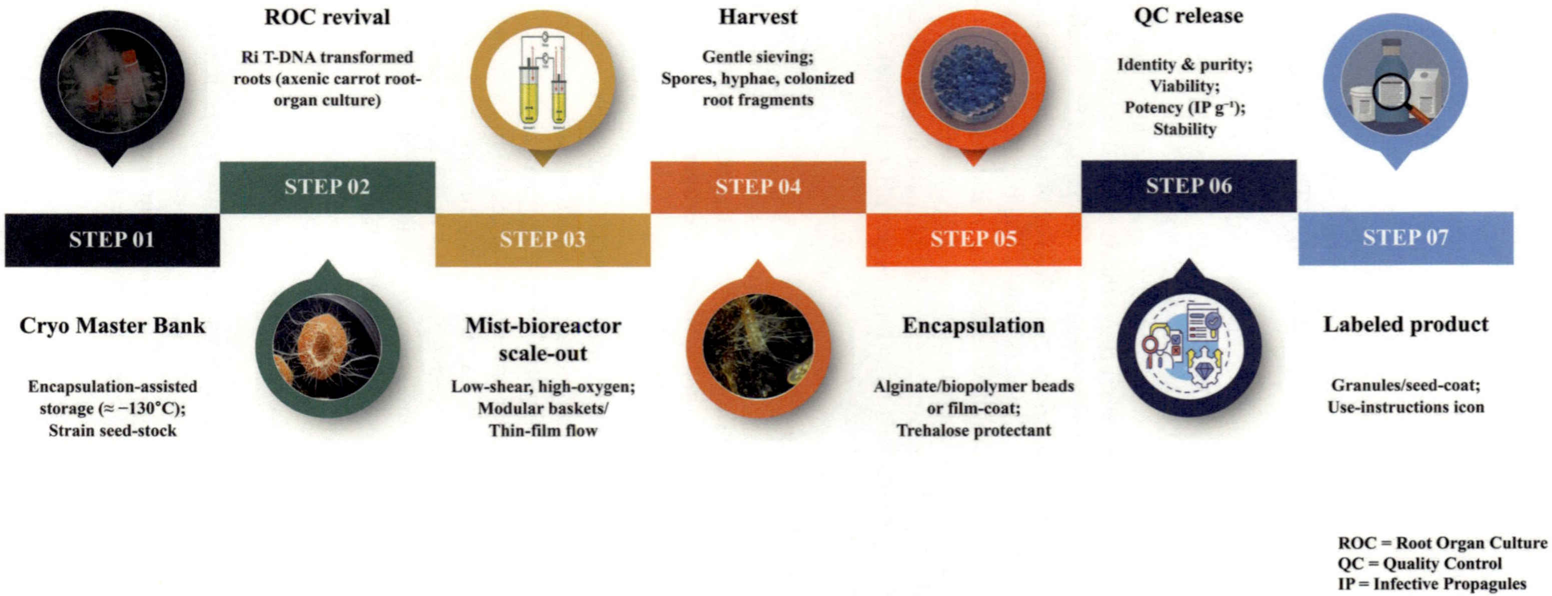

Figure 1.
Representing the process flow: cryo master bank of ROC-validated Rhizophagus and related AMF strains → revival on Agrobacterium rhizogenes (Ri T-DNA) transformed roots to establish axenic, traceable cultures → bioreactor scale-up in low-shear, high-oxygen systems (e.g., closed "mist"/gas-phase or air-lift designs with removable root baskets) to enhance volumetric productivity → harvest (disassembly, rinsing, and gentle sieving to separate spores, hyphal fragments, and colonized root pieces) → encapsulation/formulation (alginate/carrageenan beads or film-coated pellets; optional protectants) → QC release (identity/purity by morphology and then product labeling).

DOI: http://dx.doi.org/10.5772/intechopen.1013157

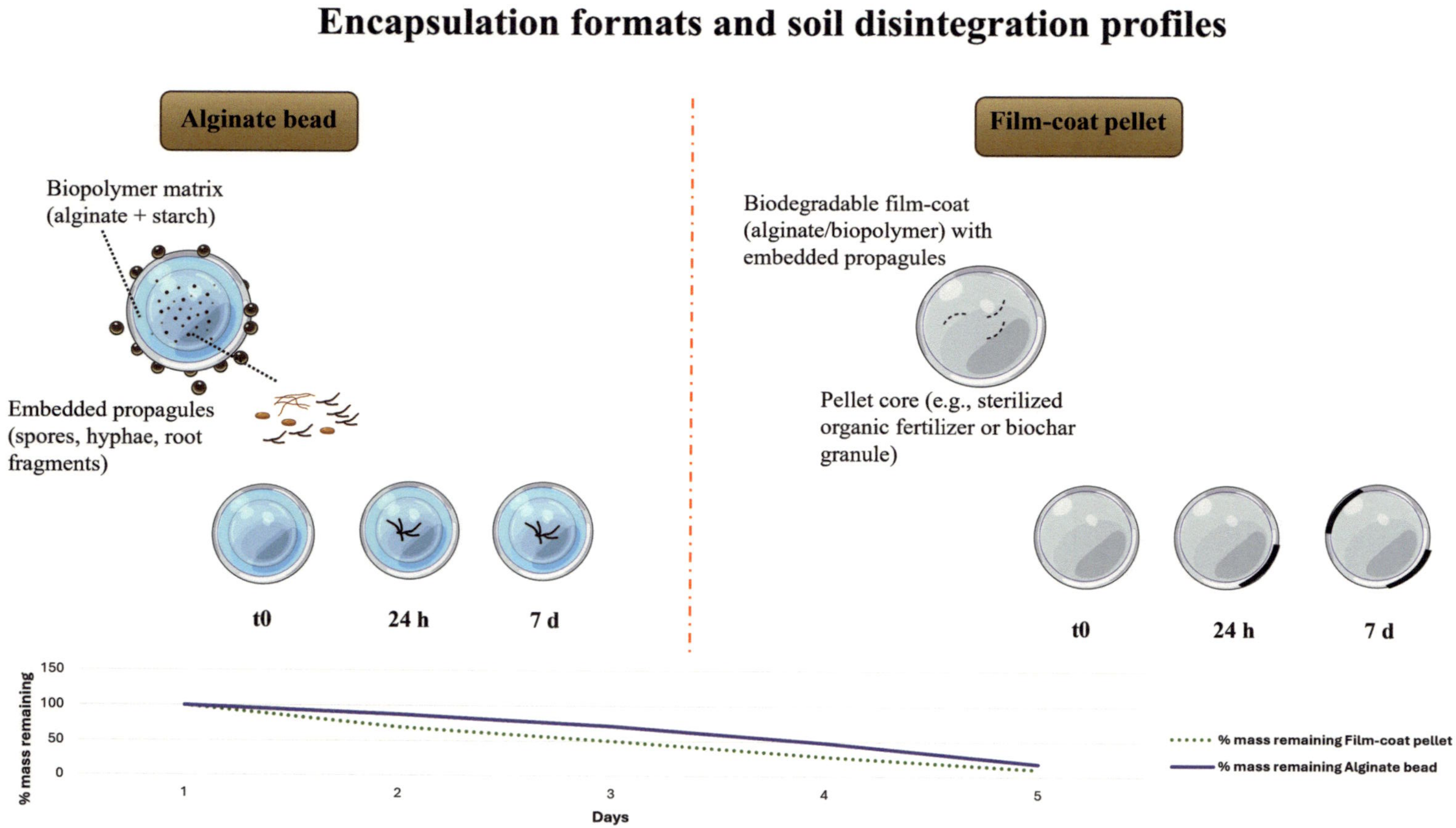

Figure 2.
Elaborates soil disintegration profile and the format of encapsulation. The figure shows a schematic comparison of the two AMF distribution forms (not to scale): (left) showing the alginate bead cross-sections. Ca-cross-linked beads absorb water and degrade over time in damp soil; time-to-release and diffusion are controlled by porosity and additives. 1B (right) illustrates the AMF propagules in thin film envelopes surrounding the organic fertilizer core (or inert granule) of the film-coated pellet. The film hydrates and breaks, releasing the propagules into the seed row. Beads typically exhibit diffusion-limited, slow release, but thin films have a quicker onset after wetting.

established ROC as the current "seed stock" platform for AMF. However, moving beyond plates introduces new engineering challenges, such as optimizing space efficiency and harvesting in a manner that does not damage fragile propagules [2–4].

2.1 Concept of bioreactor compatible with ROC

There are different types of reactors that can be used for production. These include (i) air-lift systems, (ii) packed beds, and (iii) gas-phase (mist) reactors that wet roots with recirculated medium on and off, which maximize oxygen while minimizing shear. These bioreactors are specifically intended for co-growing AMF and hairy roots (WO2019003240A1). As an alternative, one idea is to grow roots that have been modified and invaded by AMF in a continuous liquid culture using special bioreactors and harvesting schedules (WO2015028601A1). Not all geometries work better than plate ROCs, and colonization and spore spread can be completely unique. This shows how important it is to adjust hydrodynamics and supplies of O_2 for AMF biology. It is best to use low-shear, high-oxygen designs or open-work supports that prevent the roots from getting compacted and harvest-aware internals [4, 11].

2.2 Variables in the process and success measures

Scale-out depends on the type and amount of inoculum, the age and line of the roots, the composition of the medium (carbon, phosphate, micronutrients), wetting duty cycles (spray durations and intervals), and harvesting and drying processes that are set up to protect viability. It is suggested that the strength of the medium, its thickness, and the ratio of inoculum to medium have a significant effect on sporulation and colonization. These are the most important factors when moving the mixture to reactors. More than just spore numbers should be used to measure performance. We suggest a three-tier panel: (i) infective propagules (IP) using a standard most-probable -number (MPN) bioassay to check how well they work (Porter, 1979); (ii) qPCR to measure the amount of taxon-specific AMF DNA in roots and substrate to determine how they colonize and what kinds of things they make up; and (iii) spore metrics for comparing new data with old data [21, 26]. These ingredients connect the settings of the reactor to its function, not just the number of particles, and allow statistically powered technology transfer from the reactor to production.

3. Encapsulation and cryopreservation

Standardized dosage, mechanical protection during transportation and sowing, moisture buffering, and controlled release near the growing root are all made possible by the entrapment of AMF propagules. The most popular natural polymers for gel-forming carriers of soil microbial inoculants are alginate and κ-carrageenan, which can adjust the porosity, strength, and diffusion to meet desired soil disintegration timeframes [14].

Coatings provide useful field formats, particularly for AMF. Alginate film-coated organic fertilizer pellets with embedded AMF enhance early establishment, allowing for a progressive release following planting, according to case studies [12]. The following are examples of potential carriers: (i) thin films or beads of alginate,

(ii) blends of alginate and starch (stronger beads and slower diffusion), (iii) composites of alginate and bentonite, and (iv) films or coatings based on carrageenan, where a tougher film is preferred [27].

3.1 Extension of shelf life and drying protectants

During drying and storage, spores of AMF are susceptible to osmotic and thermal shocks. Disaccharides (such as trehalose and sucrose) and milk-derived solids (such as skim milk proteins and lactose) are found to be effective protectants. According to cross-domain inoculant research, these same protectants can be added to AMF bead or film formulations as long as the residues do not prevent germination [27]. The main levels of stability in practice are packaging to retain low water activity (aw), adding suitable solutes, and gently dehydrating. Microbiology of drug products provides helpful standards: as aw approaches 0.75, microbial growth usually stops. The matrices are maintained with strong moisture barriers at aw 0.30 to 0.40, which reduces metabolic activity and prolongs the shelf life of embedded bacteria. In order to improve survivability during warm-chain distribution, barrier systems using oxygen absorbers can reduce moisture stress [13].

3.2 Protecting master banks by cryopreservation

Robust master banking is achievable through encapsulation during drying and ultra-low-temperature preservation. Alginate bead encapsulation, controlled pre-drying, brief trehalose equilibration, and two-step cooling to - 130°C are among the protocols developed for in vitro-produced *Rhizophagus* isolates. After months at - 130°C, the isolates retained over 70% of potentially infectious beads, and upon revival, they recovered their life cycle [16]. Root-organ culture (ROC) recovery after thawing allows for clear expansion and certification. Modern molecular readouts (such as viability-qPCR employing intercalating dyes) and germination/colonization assays can be used to evaluate viability and integrity, supporting a GMP-like chain from the cryo master bank [17].

4. Quality control and formulation engineering

Previously, various carriers have been used for buffering moisture and ease of carrying in the industrial sector. Peat-like granules are used due to their porosity, while perlite is important for the adherence of seeds when combined with compatible stickers. An increasing number of people are looking for greener alternatives. Biochar, in particular, has a large internal surface area, a porous structure that can be adjusted, and a cation-exchange capacity, all of which work together for the delivery of nutrients and the stability of propagules. Biochar is currently being considered in formulation evaluations as a possible substitute or mixed component for bioinoculant carriers based on peat. Granules or pellets are better suited for precise dosing or mixing with fertilizers, whereas drip-like injection is also easier to place. To lessen the impact of drying and storage on spore germination, it is possible to include co-formulants such as humic compounds, sticker-spreader polymers, osmoprotectants (such as trehalose), or proteinaceous stabilizers. It is important for carriers to adhere to the microbiological and heavy metal limits imposed by the target market while maximizing infectivity and survival [21, 27].

5. Practicable QC panel

Genomic barcoding (ITS region), in conjunction with microscopy, allows for lot tracking and taxon confirmation, and screening for controlled pathogens is done per jurisdiction. The European Union Fertilizing Products Regulation (FPR) 2019/1009 states that microbial plant biostimulants (PFC 6(A)) are required to adhere to certain compositional restrictions (such as Cu and Zn caps) and fulfill the criteria for pathogen absence. Labels need to be backed by the effects that have been shown [21].

Viability: When it comes to DNA-based methods, vital stains like tetrazolium/MTT are still useful for screening spore viability. However, viability-qPCR with intercalating dyes can be adjusted for the matrix and can aid in distinguishing living from dead signals.

Effectiveness: Include all procedure information (dilution series, scoring window, confidence intervals) when reporting infective propagules (IP) using a standardized bioassay on a specific host.

Level of effectiveness: In accordance with the label and other relevant standards, declare the spores g-^{1}and/or IP g^{-}1. In the Philippines, for instance, the PNS/BAFS 183:2020 states that solid AMF inoculants must contain at least 10 spores per gram, while root inoculants must contain at least 2,300 IP g^{-1}, be nematode-free, and have a defined pH and moisture level [9].

Continued reliability: According to research on cross-domain stability, microbial proliferation is significantly reduced as a_w approaches ≤ 0.60–0.65. This means that packaging that keeps a_w low (such as barrier films with a desiccant) and prevents oxygen intrusion may prolong the shelf life of the product. Keep track of a_w targets, environmental factors, and potency as time goes on.

Some commercial AMF inoculants have been found to exhibit anomalies, such as restricted AMF viability and contamination. This integrated QC panel addresses these issues and aligns with new quality standards for mycorrhizal products. Credible labels and repeatable field performance are supported by systematic testing of identification, purity, viability, function, and stability.

There are various regulatory benchmarks that can be mapped for certification. The European Union has established regulations (EU 2019/1009) that govern the microbiological standards and labeling of microbial biostimulants with the CE mark. The Fertilizer (Control) Order, 1985 (FCO), is often cited in Indian procurement and tender specifications, as it requires mycorrhizal products to have at least 10 viable spores per gram, along with additional requirements regarding pH and IP. In the Philippines, PNS/BAFS 183:2020 establishes that solid AMF inoculants must have a minimum of 10 spores g^{-1} and root inoculants must have a minimum of 2,300 IP g^{-1}. Additionally, there are specific requirements for nematode-free status and pH windows, which are testable limits that should be reflected on certificates. All claims and repeatable field performance hinge on being explicitly aligned to these frameworks, in addition to the identity-viability-function-stability.

6. Technologies of inoculation and field deployment

It is important to choose the routes of delivery for seedlings with care to maximize root contact of the early propagule. Seed coats and films made of alginate or other biodegradable binders make it easy to dose each seed precisely and handle them easily on the farm. This can be achieved using mixers, which often contain binders and fillers that help microbes survive longer (Rocha et al., 2019). In transplant systems, plug/seedling dips work well because colonization can start in the nursery before stresses build up in the field. Encapsulated AMF can also be sent with coated pellets (for example, organic fertilizer cores covered in a biodegradable film) to make them easier to handle and place during planting. It is important to test compatibility (matrix pH, salts, fungicides) before using them together [28, 29].

Instead of counting spores, dosing should be based on the number of infective propagules per unit (IP seed^{-3} or IP cell^{-3}), which can be found using a normal Most-Probable-Number (MPN) bioassay. Active propagules are estimated by showing their presence or absence along a dilution series. This is the standard way to measure AMF infectivity, and reporting confidence intervals helps to compare products. It is important to keep the first few weeks after planting or sowing free from strong osmotic or fungal shock. This is why colonization readouts are usually done about 21 to 30 days after planting in controlled assays. Practically, there is a lot of IP in the immediate rhizosphere when planting or transplanting; being careful with water and salts, and do not using seed-treatment cocktails that do not work with each other unless you know for sure they do [30].

Some greenhouse and field studies have shown that adding MHB can help with the colonization of AMF. However, these effects depend on the situation and formulation. Both old and new reviews stress that MHB can aid mycorrhizal formation and function, but they need to be compatible with the carrier (water activity, pH, metabolites) and sometimes delivered in a staggered way (for example, bacteria on seeds and AMF in-furrow) to avoid problems [29].

7. Consortia designing

The strategy of combining AMF with the inoculum of MHB is helpful for better acquisition of nutrients for high yield. However, its impact depends on the strain type and the chemistry of the soil. The basic MHB review and newer syntheses built on this idea show how MHB supports mycorrhization by speeding up germination, moving phosphorus around, making siderophores and phytohormones, and changing the defense mechanism of plants [28].

In most cases, using an inoculum of microbes, including AMF, leads to positive and often additive effects on widely reported plant traits. However, there is still some variation, and the situation is important (for example, high soil P can cancel benefits). This makes it even more important to choose partners whose functions support each other and to test how well they work under the soil conditions of the target crop.

Formulation and transportation are as important as they are in biology. Screening for compatibility should ensure that the chosen carrier keeps the pH and water activity levels in the right ranges to survive and that the molecules do not work against each other. Bacteria can grow bigger than AMF in wet gels. In these cases, staggered delivery works well: bacteria on the seed or as a seed coat, and AMF as an in-furrow granule or gel. Seed-coating reviews stress that the type of polymer used, its loading, and the drying conditions affect the survival and release of microbes. This provides a useful way for the bacterial part to work; AMF can then be added at planting to make sure that the propagules and roots touch early [29].

Effective AMF-MHB combinations are planned and not just thrown together. You need to choose well-studied strains with different roles, make sure they do not work against each other in the carrier you want to use, use AMF dosing based on infectivity, and time the delivery to coincide with crop establishment [21, 28].

8. Economic, technical, and manufacturing considerations

Important factors for consideration included the operational and structural factors. However, the hardware involved in the bioreactor is important for control of contamination. Several of these can be trimmed by engineering decisions. In contrast to open trays, low-shear, gas-phase "mist" reactors and air-lift designs with detachable baskets minimize manual handling, facilitate modular turnover, and lower the risk of contamination. These strategies are mirrored in reactor disclosures specific to AMF that describe continuous co-culture/harvest techniques [24].

Given the significant sensitivity of unit costs to labor intensity, yield, cycle time, and uptime, a rigorous techno-economic analysis (TEA) ought to be conducted early on and repeated as data becomes more complete. Operating expenses (media, utilities, labor, and QA/QC) should be modeled, and a sensitivity analysis should be used to determine break-even drivers and inform pricing and go/no–go decisions. Despite being created for biofuels and bioproducts, these frameworks directly relate to determining the cost-effective production of AMF that could be either scaled-out (multi-unit lines) or rapidly scaled-up at early maturity [5–7].

In terms of quality and compliance, conforming to ISO/IEC 17025 (either domestically or through a third-party lab that has been accredited) shows proficiency and produces reliable and repeatable results, which are essential for export and market trust. Batch records should link validated procedures under a specified quality system to release parameters (such as spores g^{-1} and infectious propagules g^{-1}).

CE-marked microbial products must provide technical documentation and testing to demonstrate compliance with the Fertilizing Products Regulation (EU) 2019/1009, which establishes labeling guidelines, limits for contaminants and pathogens, and conformity-assessment pathways within the EU [6, 31–33].

Independent assessments have revealed inconsistent performance and viability of many commercially available AMF inoculants, highlighting the importance of rigorous quality control and clear labeling for consumer trust and repeatable field results.

9. Minimal viable process

It is better to start with an elite *Rhizophagus* isolate seed stock that has been cryopreserved and ROC-validated. Utilizing the resilience of root-organ culture (ROC) and preventing contamination drift that comes with serial culturing, revive spores or propagules on Ri T-DNA carrot roots on specified media and verify prior to any scale-out [2, 17].

Scale-out from plates to low-shear, high-O_2 reactors (such as gas-phase "mist" devices that maintain intermittently moist roots). For asepsis and reproducibility, programmed brief, recurring wetting/recirculation cycles and sterilizable hardware are used. For early production, it is important to control oxygen transmission and handling at scale [17–20].

Using the traditional wet-sieving/decanting technique modified for clean cultures, baskets are disassembled under a sterile hood at harvest, rinsed, and sieved to remove spores. In order to reduce metabolic activity without harming propagules, slow drying is important.

Propagules can be coated onto charcoal or vermiculite granules or encapsulated in alginate microbeads or films for formulation. The quality control release should consist of several layers: (i) (i) identification using ITS and taxon-specific qPCR when necessary; (ii) purity compared to regulated pathogens per destination market; (iii) potency/viability using MPN (IP) bioassay, supplemented by PMA-qPCR to distinguish between live and dead signals when necessary; and (iv) functional bioassay confirmation on an indicator host. Issue a certificate of analysis (COA) based on data particular to the lot [21, 28].

In order to reduce antagonistic interactions and position both partners in the rhizosphere, inoculation guidelines include seed coatings for nurseries, transplants, and in-furrow gels or granules for row crops.

10. Future recommendations

In order to improve the reliable production of AMF, it is essential to switch toward standard hydrodynamics to reduce heterogeneity in culture. One of the most important things to do is to compare mist and air-lift ideas to plate ROC using common measurements like infectious propagules (IP) g^{-2}, rate of contamination, and cost. Another important thing is to post head-to-head trials with clear schematics and SOPs (for example, closed mist reactors and continuous systems) [4, 11].

From "spores counted," analytics needs to move on to function-linked, species-resolved measurement. It is also important to switch to digital PCR, where absolute copy numbers make comparisons between laboratories easier. Also, make sure that extraction controls for beads, films, and granules are the same [34–36].

To make products more stable, it is better to treat them with pharmaceuticals and ensure that protective excipients (like trehalose systems) work specifically for AMF (not just PGPR). Report the shelf life using real-time/accelerated methods. Recent reviews have emphasized the need to meet stability goals for biologicals that last between 6 and 12 months.

Consortia should be planned, not just made up on the spot: Choose according to cross-domain inoculant research that has different roles (like P-solubilization, siderophores, or phytohormones), make sure they are compatible with the carrier (a_w, pH, or metabolites), and think about delivering them in stages (bacteria on seed and AMF in-furrow). It is true that context depends on things; a way to make things more reliable is to use mechanical pairing and delivery engineering.

Finally, making field data is more important. Meta-analyses show that AMF can increase yields and nutrient capture, but the results are not always consistent. Recent reviews on commercial goods have shown that they often fall short in terms of viability and performance. Multisite, QC-checked trials that confirm the identity of the product, its potency (IP g^{-2}), and its storage history are the way to go. This will help determine "where and when" AMF partnerships pay off on a large scale. Consensus and agronomic advice will happen faster with the help of public datasets and protocols that have already been recorded [10, 22].

11. Conclusion

AMF has great potential as a scalable biostimulant, depending on the quality and production optimization. Low-shear reactors that maintain symbiotic function while increasing volumetric productivity should continue to be fed from soil-free ROC, which lays the contemporary groundwork for clean seed stocks. Accelerating convergence on fit-for-purpose reactor designs can be achieved by publishing head-to-head benchmarks versus ROC plates (**Figure 1**).

Afterward, cryopreservation and encapsulation are required. The optimization of the design can be achieved with encapsulation, and master banks can be secured using encapsulation-assisted freezing (≈ - 130°C), which limits genetic drift and guarantees repeatable products throughout time. The criteria for transparent recovery should be in place before lots enter scale-out, and these tools should be standardized among the manufacturers.

Ultimately, the market's credibility can be determined by controlling its quality. From development to release, a standardized quality control panel should be implemented, including identification/purity, viability, and stability. This panel should be linked with the compliance pathways and pathogen/contaminant limits outlined in EU Regulation (EU) 2019/1009 [17–20]. Methods, labels, and COA documentation are disciplines that transcend regions [34–36].

Declaration

There is no conflict of interest. Funding is not available for this study.

Author details

Irfan Ali[1*], Taqwa Saeed[2], Rimsha Riaz[1], Rao Sohail Ahmed Khan[1], Ghulam Mustafa[1], Ahsan Iqbal[3], Tatheer Alam Naqvi[4] and Adnan Khan Niazi[1]

1 Centre of Agricultural Biochemistry and Biotechnology (CABB), University of Agriculture, Faisalabad, Pakistan

2 Faculty of Life Sciences, University of Central Punjab, Lahore, Pakistan

3 Plant Breeding and Genetics (PBG), University of Agriculture, Faisalabad, Pakistan

4 Department of Environmental Science, COMSATS University Islamabad (CUI), Abbottabad Campus, Abbottabad, Pakistan

*Address all correspondence to: irfan.cabb@uaf.edu.pk

References

[1] Adelman MJ, Morton JB. Infectivity of vesicular-arbuscular mycorrhizal fungi: Influence of host–soil diluent combinations on MPN estimates and percentage colonization. Soil Biology and Biochemistry. 1986;**18**(1):77–83. DOI: 10.1016/0038-0717(86)90106-9

[2] Bécard G, Fortin JA. Early events of vesicular-arbuscular mycorrhiza formation on Ri-T-DNA transformed roots. New Phytologist. 1988;**108**: 211–218

[3] Fortin JA, Bécard G, Declerck S, et al. Arbuscular mycorrhiza on root-organ culture: Achievements and perspectives. Canadian Journal of Botany. 2002;**80**:1–20

[4] Ijdo M, Cranenbrouck S, Declerck S. Methods for large-scale production of AMF. Mycorrhiza. 2011;**21**:1–16

[5] EU. European Union. Regulation (EU) 2019/1009 (consolidated) EUR-Lex. 2024; update 2024;1–19

[6] UK Legislation. Regulation (EU) 2019/1009—Annex III PFC 6(A): Microbial plant biostimulant (UK onshored text)

[7] O'Callaghan, M. Soil microbial inoculants for sustainable agriculture: Limitations and opportunities. Soil Use and Management. 2022;**38**:1348–1369

[8] Vassilev N, et al. Formulation of microbial inoculants by encapsulation. Frontiers in Bioengineering and Biotechnology. 2020;**8**:589

[9] Kendra S, Varga B, Gaálová-Radochová P, Bujdáková H. Practical application of PMA–qPCR assay for determination of viable cells of inter-species biofilm of Candida albicans–Staphylococcus aureus. Biology Methods and Protocols. 2024;**9** (1):bpae081. DOI: 10.1093/biomethods/bpae081

[10] Koziol L, Lubin T, Bever JD. Assessment of 23 commercially available arbuscular mycorrhizal inoculants for Sorghum bicolor and Zea mays in sterilized substrate. Applied Soil Ecology. 2024;**202**:105559. DOI: 10.1016/j.apsoil.2024.105559

[11] Schoch CL, Seifert KA, Huhndorf S, et al. Nuclear ribosomal internal transcribed spacer (ITS) region as a universal DNA barcode marker for Fungi. Proceedings of the National Academy of Sciences of the United States of America. 2012;**109** (16):6241–6246. DOI: 10.1073/pnas.1117018109

[12] Cundell AM. The role of water activity in microbial stability of non-sterile products. European Pharmaceutical Review. 2015; **20**:58–63

[13] Cichello, S. Oxygen absorbers in food preservation: A review. International Journal of Food Science and Technology. 2015;**50**:xx–xx

[14] Pitaktamrong, S., et al. Alginate film-coated fertilizer pellets embedding AMF: Handling & release. Engineering Journal (Thailand). 2018;**22**:65–79

[15] Lalaymia I, et al. Encapsulation-drying and cryostorage of AMF (-130°C): Recovery of Rhizophagus. Mycorrhiza/Fungal Biology 2012

[16] Government e-Marketplace (GeM). Specification for Mycorrhizal

Biofertilizer (as per Fertiliser Control Order, 1985) [Specification document]. 2024. Available from: https://gem.gov.in/

[17] Schoch CL, et al. Nuclear ribosomal internal transcribed spacer (ITS) region as a universal DNA barcode marker for Fungi. Proceedings of the National Academy of Sciences. 2012;**109**:6241–6246

[18] Vesper S, et al. Propidium monoazide–qPCR to distinguish viable/non-viable cells. Journal of Microbiological Methods. 2008;**72**: 180–184

[19] INVAM (University of Kansas). Most Probable Number (MPN) infectivity assay. 2025

[20] Thonar C, Erb A, Jansa J. Real-time qPCR to quantify AMF communities: Marker design and validation. Molecular Ecology Resources. 2012;**12**: xxx–xxx

[21] Rocha I, et al. Seed coating as a delivery system for beneficial microbes. Frontiers in Plant Science. 2019;**10**:1357

[22] Koziol L, Lubin T, Bever JD. Assessment of 23 mycorrhizal inoculants: Limited AMF viability, pathogen contamination. Applied Soil Ecology. 2024;**202**:105559

[23] Dos Reis GA, et al. Comprehensive review of microbial inoculants: Production, patents, and regulations. Sustainability. 2024;**16**:8720

[24] Humbird D, et al. Process design & economics for biochemical conversion of lignocellulosic biomass (TEA framework). NREL Report TP-5100-47764; 2011

[25] Bodenhausen N, et al. Relative qPCR to quantify AMF colonization in roots. Plant Methods. 2021;**17**:23

[26] Berninger T, González López Ó, Bejarano A, Preininger C, Sessitsch A. Maintenance and assessment of cell viability in formulation of non-sporulating bacterial inoculants. Microbial Biotechnology. 2018;**11** (2):277–301. DOI: 10.1111/1751-7915.12880

[27] Naz R, et al. Seed application with microbial inoculants (equipment & compatibility). In: Microbial Biotechnology for Sustainable Agriculture. Elsevier. 2022;**1**:333–366;

[28] Frey-Klett P, et al. Mycorrhiza-helper bacteria: A review. New Phytologist. 2007;**176**:22–36

[29] Nasslahsen, M., et al. Mycorrhiza-helper bacteria: Mechanisms and applications. Frontiers in Microbiology/Soil Science. 2022;**2**:1–11

[30] INVAM. Infectivity assays & mean infection percentage (MIP) method. Accessed 2025

[31] ISO/IEC 17025:2017—General requirements for the competence of testing and calibration laboratories (overview). ISO.org.

[32] A2LA. Explaining ISO/IEC 17025 competency requirements. 2024

[33] FAOLEX. Official PDF of Regulation (EU) 2019/1009. 2019

[34] Hindson BJ, et al. High-throughput droplet digital PCR

for absolute quantitation. Analytical Chemistry. 2011;**83**:8604–8610

[35] Hindson CM, et al. Absolute quantification by ddPCR vs real-time PCR. Nature Methods. 2013;**10**: 1003–1005

[36] Gobert G, Cotillard A, Fourmestraux C, Pruvost L, Miguet L, Boyer M. Droplet digital PCR improves absolute quantification of viable lactic acid bacteria in faecal samples. Journal of Microbiological Methods. 2018;**148**:64–73. DOI: 10.1016/j.mimet.2018.03.004